Environment and Natural Resources Service
Sustainable Development Department

The application of a spatial regression model to the analysis and mapping of poverty

by

Alessandra Petrucci
Professor of Statistics, University of Florence

Nicola Salvati
Project Assistant, University of Florence

Chiara Seghieri
Project Assistant, University of Florence

FOOD AND AGRICULTURE ORGANIZATION OF THE UNITED NATIONS
Rome, 2003

The designations employed and the presentation of material in this information product do not imply the expression of any opinion whatsoever on the part of the Food and Agriculture Organization of the United Nations concerning the legal or development status of any country, territory, city or area or of its authorities, or concerning the delimitation of its frontiers or boundaries.

ISBN 92-5-104973-4

Preface

Poverty mapping in developing countries has become an important tool in identifying ways to improve living standards. The methods used to generate poverty maps have come under closer scrutiny as their policy implications become more apparent. Those most commonly used until now have used econometric models to generate local indicators of poverty.

Most of these econometric models do not take into account the spatial dependence that may exist in human societies with regard to income distribution. For example, poor households are more likely to be close to other poor households than they are to be close to higher income households.

In this report, the authors use spatial regression to model more accurately the distribution of poverty across regions in Ecuador. The difference between results that are adjusted for spatial patterns and the unadjusted results is statistically significant. Although the absolute differences are not dramatic, they do provide policy planners with greater confidence that the results reflect the real situation in that country.

Although the geographic focus of this paper is on Ecuador, its main contribution is methodological, mainly the comparison of results from models that apply spatial regression techniques with those that do not.

FAO is grateful to the Government of Norway for its support to this work.

John H. Monyo
Assistant Director-General
Sustainable Development Department

The application of a spatial regression model to the analysis and mapping poverty
By Alessandra Petrucci, Nicola Salvati, Chiara Seghieri
University of Florence

64 pages, 25 Maps, 16 Tables, 1 Figure, Environment and Natural Resources Series No. 7, FAO, Rome 2003

Abstract

Although the classical econometric methods provide information on the geographic distribution of poverty, they do not take into account the spatial dependence of the data and generally they do not consider any environmental information. Therefore, methods which use spatial analysis tools are required to explore such spatial dimensions of poverty and its linkages with the environmental conditions. This study investigates an approach based on the spatial regression model, for mapping poverty in Ecuador. It also documents the use and impact of such an approach and the opportunities it offers.

Following a brief introduction, the study reviews two econometric methods that are widely used for analysing poverty and compares their results with the results obtained by the spatial regression method. The differences in numerical results were not large, but statistically significant for the Ecuador data. The study also includes a simulation study on sensitivity analysis for scenarios for public policy intervention.

Keywords: poverty mapping, spatial regression model, Ecuador, statistical estimation, GIS

This series replaces the following:
- Environment and Energy Series
- Remote Sensing Centre Series
- Agrometeorology Working Papers

A list of documents published in the above series and other information can be found at the following Web site: www.fao.org/sd

Acknowledgements

This monograph is a publication of the project *Improving Methods for Poverty and Food Insecurity Mapping and its Use at Country Level*, which is jointly implemented by FAO, the UNEP-GRID-Arendal and CGIAR centres and funded by the Government of Norway. It was prepared in collaboration with the SDRN/GIS Unit of FAO.

The authors would like to thank Professor John R. Weeks for his extensive review of the paper. His comments helped to improve the manuscript significantly. The authors would also like to thank Ergin Ataman and Barbara Huddleston for their comments and Calogero Carletto for the data he provided. Special thanks are also due to Marina Zanetti and Mario Bloise of the GIS Unit of FAO for their help with the GIS mapping and analysis; to Julian Plummer for editing the manuscript and to Chiara Caproni for the graphics and page layout of the paper.

Contents

List of tables

List of figures

List of acronyms

ECV Encuesta Condiciones de Vida (survey of living standards)
EDA Exploratory data analysis
GIS Geographical information system
HIE Household income and expenditure
LSMS Living Standard Measurement Surveys (World Bank project)
MAUP Modifiable areal unit problem
MPE Maximum pseudo-likelihood estimator

Chapter 1. **Introduction**

Poverty maps are important tools that provide information on the spatial distribution of poverty within a country. They are used to affect various kinds of decisions, ranging from poverty alleviation programmes to emergency response and food aid.

However, the use of poverty maps alone does not furnish an estimate of the causal linkage between poverty and the variables that influence it; such maps furnish only 'visual' advice. For this reason, researchers usually look for the possible existence of empirical relationships between poverty and socio-economic indicators. They make use of statistical methods such as the econometric model that combines census and survey data as applied in South Africa and Ecuador (Hentschel *et al.*, 2000).

Another characteristic of poverty map studies is that to date they have not taken account of the geographical components (location) and environmental data that may have an important impact on research results.

Environmental degradation contributes to poverty through worsened health and by constraining the productivity of those resources on which the poor rely. Moreover, poverty restricts the poor to acting in ways that harm the environment. Poverty is often concentrated in environmentally fragile ecological zones where communities face and contribute to different kinds of environmental degradation. In addition, demographic factors can be involved in complex ways (high population growth rates are associated with poverty) and exacerbate problems of environmental degradation.

In the social sciences, spatial contiguity in social and economic variables is a consequence of the instincts of individuals and of the patterns of behaviour and economic constraints that taken together help bind social space into recognizable structures. In a village or urban community, many of the households may have similar sources of income, and all households are affected by the same agroclimatic and geographic conditions. They also have other circumstances in common including road conditions, availability of public facilities for services such as health, water supply and education. Hence, it is reasonable to suppose that households living in the same area tend to act in similar ways and to influence one another.

In many countries, poverty and food security are highly heterogeneous phenomena showing a wide spatial variability. Large differences between the standard of living of populations in different geographical areas are common in both developed and developing countries. Spatial heterogeneity between areas can be introduced for a variety of reasons, including differences in agroclimatic conditions, geographic conditions (particularly access to main urban centres and markets), the presence of natural resources (particularly water for irrigation), other non-physical conditions (historical, ethnical, etc.) and facets of public policy (Jalan and Ravallion, 1998).

However, it is often difficult to measure heterogeneity in poverty and food security correctly using conventional analytical tools. The key problem is one of obtaining data that permit the measurement of poverty and food security at a level of disaggregation that is sufficient to capture the heterogeneity related to spatial variability.

Chapter 2. **Methods for mapping poverty at subnational levels**

Various methods for measuring the spatial location of the poor have been put forward in the literature and in practice, and most of them are under continuing development. Each of these methods has different data requirements and implementation costs.

This chapter provides a brief description of two widely known methods proposed by staff at the World Bank and researchers of other international agencies. One method is based on unit-level data and the other takes into account community-level data.

Poverty mapping based on household survey and census data

Poverty mapping based on household survey and census data is a method that has been used in a series of World Bank studies (Lanjouw, 1998; Hentschel *et al.*, 2000; Elbers *et al.*, 2001; and Deichmann, 1999).

This method, henceforward referred to as the Lanjouw *et al.* method, requires a minimum of two sets of data: household-level census data and a representative household survey corresponding approximately to the same time period as the census. It estimates the incidence of poverty in administrative areas (districts, subdistricts, counties, etc.) on the basis of the data of the household income and expenditure (HIE) survey and the population census. The measures of poverty estimated with this method are based on estimates of the individuals' (or households') incomes and these are therefore the common poverty measures that are used in welfare analysis. To determine these estimates, the method uses the HIE survey data and the data of the population census to estimate the per capita incomes of all the individuals in the country or the province (in other studies the dependent variable is the probability that the income per person falls below the poverty line). Separate estimates are made for individuals in each administrative area and separate measures of poverty for each area are calculated with these estimates.

This approach combines an intensive survey containing information on consumption for a sample of areas and households with individual census data (or an extensive survey). The household survey data are used to estimate a consumption equation:

$$y_i = f(x_i)$$

where y_i is income/consumption expenditures per capita for household i, and x_i is a vector of household characteristics. The choice of the indicators must be made so that the explanatory power of the vector $(x_{i1}, \ldots, x_{im})$ is sufficiently high. At the same time, only variables that are available both in the household survey and in the census data can be selected. The equation is used to predict the probability that households in a given area are poor for all the areas in the country, using the entire census data. A more detailed description of the specific steps in this analysis is given below.

In the first step, the data of the HIE survey are used in an econometric analysis to estimate the functional relations between the level of the household's income and key household and community characteristics. In the second step, these functional relations are used to estimate the level of income for all individuals/households in the country on the basis of the actual data from the population census. In the third step, these income estimates are used to determine the measure of poverty.

The practical application of this concept presents a number of econometric and computational challenges, including issues relating to the large size of census data sets, non-normality, spatial autocorrelation, and heteroscedasticity. Elbers *et al.* (2001) discuss these aspects in detail. In particular, the implementation of this method requires a huge database created by a well-designed and large-dimension survey on consumption/expenditure (HIE) and the census data. One quality of this methodology is the relative ease of checking the reliability of estimates that are built into the programs the World Bank provides to national poverty-mapping analysts. The size of standard errors in these estimates depends largely on the degree of disaggregation sought and the explanatory power of the exogenous variables in the first-stage model.

Poverty mapping based on household survey and area indicators

Another estimation method involves the use of average values from disaggregated geographical units, such as communities or small towns, instead of household-level data (Bigman *et al.*, 2000). In this study, this method is referred to as the Bigman *et al.* method.

The Bigman *et al.* method has the advantage of less stringent data requirements. National statistical agencies may be more likely to release community averages upon request. Indeed, these data may be published. Apart from the difference in scale of the predictive model, the two estimation methods follow essentially the same steps.

The Bigman *et al.* method estimates an index of poverty or vulnerability for each community (district, subdistrict or county). This index is the probability that per capita consumption expenditures of households in a given community fall below the poverty line. In this case, the vulnerability is the incidence of poverty in the community. With this method, the index is estimated by using data from a wide range of sources that provide detailed information on the community and the area in which it is located. These sources can include: geographical information system (GIS) coverages that provide data on the geographic, topographic and climate conditions in the area; road maps that give data on the quality of the road infrastructure, the distance to the nearest market, etc.; the agricultural census that provides data on the crops grown in that area, the use of irrigation, fertilizers and pesticides in that area, etc.; and various surveys (such as the demographic and health survey) that provide data on the demographic and socio-economic characteristics of the population in these areas, etc. Data on community characteristics obtained from the population census can also be used.

This method combines the household survey with information on the areas in which the individuals reside (rather than on their personal data). For the analysis, information on the areas is collected from a wide variety of sources. The first step in calculating the vulnerability index for each community is to use the individual household survey data in order to estimate an income or consumption equation:

$$y_{ia} = f(x_i, z_a)$$

where y_{ia} is the income or consumption expenditures of individual i in area a, $(x_{i1}, \ldots, x_{im})$ is a vector of the household's personal and family characteristics, and $(z_{i1}, \ldots, z_{ak})$ is a vector of the area characteristics. From this logit analysis, one can derive estimates for the probability that an individual in area a is poor as:

$$P_a = g(\bar{x}_a, z_a)$$

where $\bar{x}_a$ is the average values of the individual characteristics ($x_{ia1}, \ldots, x_{iam}$) for all the individuals in each of the areas *a*. These relations can be used to predict the incidence of poverty P_a in all small 'areas' of the country (subdistricts, counties, etc.) on the basis of the information on the area characteristics ($\bar{x}_a$, z_a), which can be obtained from a wide variety of sources.

While easier access to data makes this method attractive, the error associated with estimation for units of different sizes (in terms of population) has not been investigated thoroughly. Thus, it is not clear what the trade-offs are between ease of access and statistical reliability.

Chapter 3. **The data**

The first source of data considered for the purposes of this study is the Encuesta Condiciones de Vida (ECV) database. This database stems from a large and comprehensive survey conducted on Ecuador in 1995 that forms part of the World Bank's Living Standard Measurement Surveys (LSMS) project that started in 1980. The survey was administered to a sample of about 5 800 nationally representative households (this study makes use of a reduced sample of 5 630 households because data were missing for some variables). It collected data on all dimensions of household well-being and socio-economic characteristics including highly disaggregated data on household consumption expenditures. The survey design incorporated both clustering and stratification on the basis of the country's three main agroclimatic zones and rural-urban breakdown. It also oversampled Ecuador's two main cities: Quito and Guayaquil.

According to the ECV sampling design, the sample employed in this study is representative of the main agroclimatic zones. The sample size is too small to allow an estimation of the incidence of poverty at the level of provinces, counties and *parroquias* (municipalities). On the basis of this survey, if traditional mapping methods of spatial distribution of poverty were applied, the only working level should be the main regions in Ecuador (first level). However, by using data from another source, the INFOPLAN database, and aggregating the two databases at the common level of the county, it was possible to map the spatial distribution of poverty at county level.

INFOPLAN is an atlas that collects about 104 variables from the "Census of population and households" (INEC) conducted in Ecuador in 1990. It provides a wide variety of information on the demographic, socio-economic and geographical characteristics of the areas, and the data are available for many geographical area levels (from regions to *parroquias*). However, it does not contain income or consumption expenditure information for each household. Although the 1995 ECV data were collected five years after the census, the 1990-95 period was one of relatively slow growth and low inflation in Ecuador, so it is reasonable to assume that there was relatively little change.

Furthermore, the household living standards in the available counties (Table 6 in Chapter 5) are not georeferenced: the location of the respondent household was identified only by the county of residence and the type (rural or urban) of living area. This problem was overcome by locating each family randomly (assuming a uniform distribution) in the county but taking into account the type of living area. Moreover, in order to understand the relationship between poverty and environment, the study also considered some environmental variables at the county level, e.g. cereal production, amount of arable land, and the distance of the households from the main roads (data provided by FAO/SDRN GIS).

Finally, all the data from the three sources of information (ECV, INFOPLAN and FAO) were arranged in a GIS for managing the spatial dimension. The FAO and INFOPLAN data were already organized as GIS data and could be overlaid by merging the information contained in the layers. From the ECV data, a point coverage was created taking into account the geographical constraints.

Table 1. **Estimated percentage of poor people in Ecuador: a comparison between two methods**

Area	**Method**		
	Lanjouw *et al.* (1994)	**Lanjouw *et al.* (1995)**	**Bigman *et al.* (1995)**
Costa	**40** *	**54**	**46**
Urban	35	43	33
Rural	52	75	75
Sierra	**42** *	**58**	**42**
Urban	33	42	17
Rural	53	78	69
Oriente	**60** *	**65**	**69**
Urban	25	47	36
Rural	67	70	75
Ecuador	**41** *	**56**	**45**
Urban	34 *	42	26
Rural	53 *	76	72

* *Outputs of the computation.*

Numerical comparison on the Ecuador data

Table 1 reports the outcomes from the comparison between the two methods presented above using the above data. The results for the Lanjouw *et al.* method were taken directly from the authors' papers (Hentschel and Lanjouw, 1996; Hentschel *et al.*, 2000), while the outcomes from the Bigman *et al.* method were computed using ad hoc programs (Annex C) with the 1995 data.

In particular, Table 1 shows:

- The percentage of poor individuals increases from 1994 to 1995 under the Lanjouw *et al.* method.
- In some rural areas, the two methods produce approximately analogous results for 1995 data. However, there is a large discrepancy in the percentage of poor in urban areas.
- The differences between rural and urban areas increase for the Sierra region.

Chapter 4. **Spatial analysis**

In general, poverty maps do not measure casual linkages between variables. Hence, in order to describe these casual relationships, it is necessary to use an appropriate statistical analysis.

In particular, this study applies a spatial analysis to determine those variables that affect household poverty and to estimate the number of poor people in the target areas. This type of analysis is based on the assumption that measured geographic variables often exhibit properties of spatial dependency (the tendency of the same variables measured in locations in close proximity to be related) and spatial heterogeneity (non-stationarity of most geographic processes, meaning that global parameters do not well reflect processes occurring at a particular location). While traditional statistical techniques have treated these two last features as nuisances, spatial statistics considers them explicitly.

From the methodological point of view, the spatial analysis is based on five steps:

- Step I. The spatial estimation of the impact of location characteristics of the areas in which the households reside is used to calculate the probability that these households are poor. The household data from the ECV and the community data from INFOPLAN are employed in order to determine the variables that best explain household consumption and poverty (Table 2).
- Step II. Basic exploratory data analysis (EDA) techniques are applied and the spatial neighbourhood structure is defined in order to test the presence of spatial autocorrelation among the observed values.
- Step III. The incidence of poverty in all the target areas (counties) in the country (Ecuador) is estimated on the basis of their location-specific characteristics (Table 3) and on the relationship estimated in Step I.
- Step IV. The first validation of the estimation is executed comparing the incidence of poverty at the county level established by the spatial econometric estimation with the level of poverty in the county computed from the ECV survey data. This validation is based on the ranking of counties.
- Step V. The second validation of the predictions is performed, aggregating the results at the province level and comparing this incidence of poverty with the Bigman *et al.* method and the Lanjouw *et al.* method.

In accordance with other similar studies, consumption is assumed to be the welfare indicator at the household level, and the headcount index is used as the measure of poverty.

Spatial autocorrelation

Spatial autocorrelation is a property of spatial data that exists whenever there is a systematic pattern in the values recorded at locations in a map. In particular, where high values of a variable at one locality are associated with high values at neighbouring localities, the spatial autocorrelation is positive, and where high values correspond to low values in the adjacent localities, the spatial autocorrelation is negative.

In order to detect the spatial pattern (spatial association and spatial autocorrelation), some standard global and new local spatial statistics have been developed. These include Moran's I, Geary's C,

TABLE 2. **Descriptive statistics on variables used in the estimation**

Aggregation level	Variable	Type	Urban			Rural			Data source
			Mean	Standard error	No. observations	Mean	Standard error	No. observations	
Household	Percentage adults illiterate in household	Numeric	0.09	0.18	3366	0.10	0.19	2264	ECV
Household	Percentage persons with diploma[1]	Numeric	0.06	0.16	3366	0.007	0.05	2264	ECV
Household	Adequate home	Dummy	0.93	0.25	3366	0.74	0.43	2264	ECV
Household	Home with drinking water	Dummy	0.79	0.40	3366	0.27	0.44	2264	ECV
Household	Home with adequate toilet	Dummy	0.94	0.23	3366	0.65	0.47	2264	ECV
Household	Home with adequate wall	Dummy	0.78	0.41	3366	0.40	0.49	2264	ECV
Household	Home with electricity public network	Dummy	0.97	0.16	3366	0.74	0.43	2264	ECV
Household	Waste: collection by truck	Dummy	0.76	0.42	3366	0.13	0.33	2264	ECV
Household	Persons per room	Numeric	1.83	1.37	3366	2.30	1.73	2 264	ECV
County	Population	Numeric	56 541.00	194 463.55	99	18 947.00	20 279.47	210	INFOPLAN
County	Mortality rate (‰)	Numeric	47.01	11.31	99	66.42	14.81	210	INFOPLAN
County	Number of babies	Numeric	1 348	4 583.94	99	502	504.09	210	INFOPLAN
County	People < 5 km from road[3]	Numeric	32 267.62	99955.70	–	32 267.62	99955.70	–	FAO/SDRN GIS
County	People 5-15 km from road[3]	Numeric	5 542.75	14 347.99	–	5 542.75	14 347.99	–	FAO/SDRN GIS
County	People > 15 km from road[3]	Numeric	1 056.43	3 696.94	–	1 056.43	3 696.94	–	FAO/SDRN GIS
County	County area (km^2)	Numeric	1 296.01	2 199.57	99	1 176.51	1 938.78	210	FAO/SDRN GIS
County	Cereal production coefficient	Numeric	205.20	416.94	99	140.30	302.30	210	FAO/SDRN GIS
County	Protected area	Dummy	0.26	0.44	99	0.28	0.45	210	FAO/SDRN GIS
County	> 35% irrigation area	Dummy	0.35	0.48	99	0.22	0.42	210	FAO/SDRN GIS
County	Closed forest	Dummy	0.29	0.46	99	0.33	0.47	210	FAO/SDRN GIS
County	Arable land[2]	Factor	–	–	99	–	–	210	FAO/SDRN GIS
Province	Erosion	Factor	–	–	21	–	–	21	INFOPLAN
Province	Climate	Factor	–	–	21	–	–	21	INFOPLAN
Province	Flooding area	Dummy	0.52	0.51	21	0.52	0.51	21	INFOPLAN
Province	Volcano area	Dummy	0.38	0.49	21	0.38	0.49	21	INFOPLAN

(1) *Secondary school.*

(2) *Arable land for each counties is obtained by first multiplying the weighted average area of the county by the percentage of arable land in the same county. The result is transformed as a factor which takes a value of 0 if the number obtained is < 30%, 1 if between 30 and 60%, 2 if > 60%.*

(3) *People < 5 km from road, People 5–15 km from road, People > 15 km from road are calculated on all the counties without distinction between urban and rural county.*

TABLE 3. **Descriptive statistics on variables used in the prediction**

Aggregation level	Variable	Type	Urban			Rural			Data source
			Mean	Standard error	No. observations	Mean	Standard error	No. observations	
County	Percentage adults illiterate in county	Numeric	0.09	0.04	99	0.19	0.08	210	INFOPLAN
County	Percentage persons with diploma[(1)]	Numeric	0.13	0.06	99	0.04	0.02	210	INFOPLAN
County	Percentage of Adequate home	Numeric	0.91	0.06	99	0.69	0.19	210	INFOPLAN
County	Percentage of home with drinking water	Numeric	0.51	0.24	99	0.13	0.10	210	INFOPLAN
County	Percentage of home with adequate toilet	Numeric	0.63	0.12	99	0.20	0.12	210	INFOPLAN
County	Percentage of home with adequate wall	Numeric	0.66	0.15	99	0.29	0.20	210	INFOPLAN
County	Percentage of homes w. public electricity network	Numeric	0.91	0.05	99	0.48	0.22	210	INFOPLAN
County	Waste: collection by truck (%)	Numeric	0.63	0.20	99	0.09	0.11	210	INFOPLAN
County	Persons per room	Numeric	2.23	0.43	99	2.60	0.47	210	INFOPLAN
County	Population	Numeric	5 6541	194 463.55	99	1 8947	20 279.47	210	INFOPLAN
County	Mortality rate (‰)	Numeric	47.01	11.31	99	66.42	14.81	210	INFOPLAN
County	Number of babies	Numeric	1 348	4 583.942	99	502	504.09	210	INFOPLAN
County	People < 5 km from road[(3)]	Numeric	32 267.62	99955.70	–	32 267.62	99955.70	–	FAO/SDRN GIS
County	People 5-15 km from road[(3)]	Numeric	5 542.75	14 347.99	–	5 542.75	14 347.99	–	FAO/SDRN GIS
County	People > 15 km from road[(3)]	Numeric	1 056.43	3 696.94	–	1 056.43	3 696.94	–	FAO/SDRN GIS
County	County area (km^2)	Numeric	1 296.01	2 199.57	99	1 176.51	1 938.78	210	FAO/SDRN GIS
County	Cereal production coefficient	Numeric	205.20	416.94	99	140.30	302.30	210	FAO/SDRN GIS
County	Protected area	Dummy	0.26	0.44	99	0.28	0.45	210	FAO/SDRN GIS
County	> 35% irrigation area	Dummy	0.35	0.48	99	0.22	0.42	210	FAO/SDRN GIS
County	Closed forest	Dummy	0.29	0.46	99	0.33	0.47	210	FAO/SDRN GIS
County	Arable land[(2)]	Factor	–	–	99	–	–	210	FAO/SDRN GIS
Province	Erosion	Factor	–	–	21	–	–	21	INFOPLAN
Province	Climate	Factor	–	–	21	–	–	21	INFOPLAN
Province	Flooding area	Dummy	0.52	0.51	21	0.52	0.51	21	INFOPLAN
Province	Volcano area	Dummy	0.38	0.49	21	0.38	0.49	21	INFOPLAN

(1) Secondary school.
(2) Arable land for each counties is obtained by first multiplying the weighted average area of the county by the percentage of arable land in the same county. The result is transformed as a factor which takes a value of 0 if the number obtained is < 30%, 1 if between 30 and 60%, 2 if > 60%.
(3) People < 5 km from road, People 5–15 km from road, People > 15 km from road are calculated on all the counties without distinction between urban and rural county.

G statistics (Getis, 1992), LISA (Anselin, 1995) and GLISA (Bao and Henry, 1996). All these spatial analytical techniques have two aspects in common. First, they start from the assumption of a spatially random distribution of data. Second, the spatial pattern, spatial structure, and form of spatial dependence are typically derived from the data (Bao, 1999).

The first measure of spatial autocorrelation was introduced by Moran (1950). The index is analogous to the conventional correlation coefficient, and its values range from 1 (strong positive spatial autocorrelation) to -1 (strong negative spatial autocorrelation). It is often used to measure the spatial autocorrelation of ordinal, interval or ratio data. Moran's I is defined by:

$$I(d) = \frac{\sum_{i}^{n} \sum_{j \neq i}^{n} w_{ij}(x_i - \bar{x})(x_j - \bar{x})}{S^2 \sum_{i}^{n} \sum_{j \neq i}^{n} w_{ij}} \tag{1}$$

where $S^2 = \frac{1}{n} \sum_{i}^{n} (x_i - \bar{x})^2$, x_i denotes the observed value of population at location i, $\bar{x}$ is the average of the x_i over the n locations, and w_{ij} is the spatial weight measure of contiguity and is defined as 1 if location i is contiguous to location j and 0 otherwise.

The choice of weights w_{ij} between neighbours is a crucial step in the analysis. There are many ways to assign neighbour weights, and the choice depends on the type of spatial application and on the research question. This specification requires *a priori* knowledge of the range and intensity of the spatial covariance between regions. Common methods include row standardization, length of common boundary and distance functions.

With regard to the weighting procedure, the use of generalized weighting matrix *W*, as opposed to a binary connection matrix, allows a set of weights to be chosen which are deemed appropriate from prior consideration. This matrix *W* offers great flexibility in defining the structure of the county system, and it permits items such as natural barriers and county size to be taken into account. It is important to note that the elements of *W* are non-stochastic and exogenous to the model. Typically, they are based on the geographic arrangement of the observations or contiguity. Weights are non-zero when two locations share a common boundary or are within a given distance of each other. However, this notation is general and alternative specifications of the spatial weights can be based on distance decay (inverse distance). In this study, the neighbourhood structure is based on inverse Euclidean distance (Anselin, 1992).

The expected value and variance of Moran's I for a sample of size *n* could be calculated according to the assumed pattern of the spatial data distribution (Cliff and Ord, 1981).

For the assumption of a normal distribution:

$$E_n(I) = -\frac{1}{(n-1)} \tag{2}$$

$$VAR_n(I) = \frac{n^2 w_1 - n w_2 + 3 w_0^2}{w_0^2 (n^2 - 1)} - E_n^2(I) \tag{3}$$

For the assumption of random distribution:

$$E_r(I) = -\frac{1}{(n-1)} \tag{4}$$

$$VAR_r(I) = \frac{n\left((n^2-3n+3)w_1 - nw_2 + 3w_0^2\right) - K_2\left((n^2-n)w_1 - 2nw_2 + 6w_0^2\right)}{w_0^2(n-1)(n-2)(n-3)} - E_r^2(I) \tag{5}$$

where: $K_2 = \frac{n\sum_i^n (x_i-\bar{x})^4}{\left(\sum_i^n (x_i-\bar{x})^2\right)^2}$, $w_0 = \sum_i^n\sum_j^n w_{ij}$, $w_1 = \frac{1}{2}\sum_i^n\sum_j^n (w_{ij}+w_{ji})^2$, $w_2 = \sum_i^n (w_{i.}+w_{.i})^2$,

$w_{i.}$ and $w_{.i}$ are the sum of the row i and column i of the weight matrix respectively.

The test of the null hypothesis that there is no spatial autocorrelation between observed values over the n locations can be conducted on the basis of the standardized statistics as follows:

$$Z(d) = \frac{I(d) - E(I)}{\sqrt{VAR(I)}} \tag{6}$$

Moran's I is significant and positive when the observed values of locations within a certain distance (d) tend to be similar, negative when they tend to be dissimilar, and approximately zero when the observed values are arranged randomly and independently over space.

Another index for testing the presence of spatial autocorrelation in the data is Geary's C. It uses the sum of squared differences between pairs of data values as a measure of covariation. The formula of Geary's C is:

$$C(d) = \frac{(n-1)\sum_i^n\sum_{j\neq i}^n w_{ij}(x_i-x_j)^2}{nS^2\sum_i^n\sum_{j\neq i}^n w_{ij}} \tag{7}$$

This index ranges between 0 and 2. Positive spatial autocorrelation is found with values ranging from 0 to 1 and negative spatial autocorrelation is found between 1 and 2.

For the assumption of a normal distribution:

$$E_n(C) = 1 \tag{8}$$

$$VAR_n(C) = \frac{(2w_1+w_2)(n-1) - 4w_0^2}{2w_0^2(n+1)} \tag{9}$$

For the assumption of random distribution:

$$E_r(C) = 1 \tag{10}$$

$$VAR_r(C)=\frac{(n-1)w_1(n^2-3n+3-K_2(n-1))}{w_0^2 n(n-2)(n-3)}+\frac{(n^2-3-K_2(n-1)^2)}{n(n-2)(n-3)}-\frac{(n-1)w_2(n^2+3n-6-K_2(n^2-n+2))}{4n(n-2)(n-3)w_0^2} \quad (11)$$

The significance of Geary's C is tested identically to that for Moran's I.

The results of spatial autocorrelation tests should be used with caution. First, the choice of neighbours and their respective weights determines the values of Moran and Geary statistics. A non-significant result indicates that there is no significant spatial autocorrelation given the neighbourhood structure provided. Second, a significant positive autocorrelation could be caused by a spatial pattern in the data not specified by the statistical model. The following section shows how this model misspecification can be controlled by incorporating a spatial weights matrix into the statistical model.

Generalized spatial linear models

As a special case, generalized spatial linear models include spatial linear regression and analysis of variance models, spatial logit and probit models for binary responses, loglinear models and multinomial response models for counts.

Let c_i denote the level of consumption per household, z denote the poverty line, and $s_i = c_i / z$ be the normalized welfare indicator per household. The household poverty indicator is determined by the normalized welfare function as follows:

$$y_i = 1 \text{ if } \ln s_i < 0$$

$$y_i = 0 \text{ if } \ln s_i \geq 0.$$

The households are observed in n sites that form a subset S of the space. Each point (household) i has a binary response y_i and a vector k x 1 of covariates x_i. The responses constitute a map $Y = (y_i)_{i=1}^{n}$.

The regression model is called autologistic and states the conditional probability p_i that y_i is equal to 1, given all other site values y_j $(j \neq i)$:

$$p_i = \Pr(y_i = 1 | y_j, j \neq i) = \Pr(y_i = 1 | y_j, j \in N(i)) = \Phi(\beta_0 + \beta_i^H x_i^H + \beta_i^C x_i^C + \gamma y_i^*) \quad (12)$$

where $N(i)$ is the neighbour set of site i according to a neighbourhood structure and y_i^* is the sum of the values of the dependent variable of the neighbours of the site i, that is:

$$y_i^* = \sum_{j=1}^{n} y_j I(i \cong j) = \sum_{j:i \cong j} y_j \quad (13)$$

where $i \cong j$ denotes that the households i and j are neighbours.

This kind of model then takes into account the spatial distribution of the welfare indicator, incorporating the neighbourhood structure in the model as another parameter to estimate.

In the model, X^H is the vector of explanatory variables that describe the household characteristics, X^C is the vector of explanatory variables that describe the characteristics of the area in which the households reside, and Φ is a cumulative distribution function that is standard normal in the case of probit regression.

For a given poverty line and a given set of observation on X^H and X^C, the estimates of β^H, β^C and γ can be obtained by the maximum pseudo-likelihood method. Besag (1975) has demonstrated that the pseudo-likelihood method produces consistent parameter estimates under regular conditions.

Given the above generalized linear model, a maximum pseudo-likelihood estimator (MPE) for the unknown parameter vector $\theta = \{\beta_0, \beta^H, \beta^C, \gamma\}$ will be defined as the vector $\hat{\theta}$ that maximizes the pseudo-likelihood function:

$$\prod_{i=1}^{n} \Pr(y_i \mid y_j, j \neq i) = \prod_{i=1}^{n} p_i^{y_i} (1 - p_i)^{y_i} \qquad (14)$$

As a result, the function in Equation 14 is not a full likelihood. An analytical form of the full likelihood is intractable for this problem because there is generally an unknown normalizing function.

Therefore, the pseudo-likelihood estimation procedure proposed is an intuitively plausible method that avoids the technical difficulties of the full maximum likelihood approach. A drawback of the method is that its sampling properties have not been studied as extensively as those of the full maximum likelihood estimators.

Besag (1977) discusses the consistency and efficiency of pseudo-likelihood estimation for simple spatial Gaussian schemes. Strauss and Ikeda (1990) have shown that, for a logit model, maximization of Equation 14 is equivalent to a maximum likelihood fit for a logit regression model with independent observations y_i. Consequently, estimates can be obtained by using an iteratively reweighted least squares procedure.

Therefore, any standard logistic regression routine can be used to obtain MPEs of the parameters. However, the standard errors of the estimated parameters calculated by the standard programs are not directly applicable because they are based on the assumption of independence of the observations.

The next step is the estimation of the incidence of poverty in all counties. These estimates are made on the basis of the relationship between the area characteristics and the probability that households residing in these areas are poor. The probability that households in a given county are poor is estimated only on the basis of the area characteristics:

$$\pi_C = \Phi^{-1}(X^C \beta^H + X^C \beta^C) \qquad (15)$$

where β^H and β^C are the coefficients from Equation 12 and π_C is the probability that a household drawn from a certain county is poor. Therefore, the parameter estimates from the regression are applied to the census data in order to obtain an imputed value for π_C, the percentage of poor households in a county. In this way, the poor households in all the counties are estimated. Finally, using the information on household size, the probability of a household being poor can be extended to the probability of an individual being poor.

Chapter 5. **Empirical estimation and predictions**

In the analysis of poverty in Ecuador, the spatial autocorrelation of household consumption was calculated for the total sample (5 630 households), the rural households (2 264) and the urban households (3 366). Table 4 shows that there is a large and significant spatial autocorrelation in rural household consumption, whereas it is small but significant for the total sample and urban household consumption.

In order to account for the presence of significant spatial autocorrelation in the data, three spatial probit models were constructed: for the total sample and for the subsamples of the rural and urban households. The coefficient values, their *z*-value in all three models, were analysed and it was verified that there were no large differences. Hence, it was decided to use only one model: the total sample model.

Table 5 gives the estimated coefficients of the spatial probit regression that estimates the probability that the household is poor as a function of various households' characteristics, various characteristics of the area in which the household resides and of a component (y^*) standing for the spatial dimension. Table 5 reports the standard errors. Even if they do not have a theoretical meaning, they can have a descriptive value and provide some general information. However, some studies where the standard errors are computed both by standard statistical packages and bootstrap simulation techniques remark on the comparability of the results.

In both rural and urban areas, the household variables that have a significant relationship to a household's probability of being poor are the adult literacy rates (if the components of the household have a diploma, the household's probability of being poor decreases). The same effect is related to the adequacy of the house and to the presence of a waste collection truck (if the household lives in a house with solid walls and a toilet and there is the regular presence of a waste collection truck, the household's probability of being poor decreases). Moreover, as the number of persons living in a single room of the house increases, the household's probability of being poor also increases.

TABLE 4. **Spatial autocorrelation**

Spatial correlation estimate (total sample)
Statistic = "Moran"
Sampling = "free"
Correlation = 0.177
Null hypothesis: no spatial autocorrelation (rejection)
Spatial correlation estimate (rural sample)
Statistic = "Moran"
Sampling = "free"
Correlation = 0.651
Null hypothesis: no spatial autocorrelation (rejection)
Spatial correlation estimate (urban sample)
Statistic = "Moran"
Sampling = "free"
Correlation = 0.0532
Null hypothesis: no spatial autocorrelation (rejection)

All the variables at the county level have a significant correlation with poverty. In particular, households living in counties with a high population density and high mortality rates have high probabilities of being poor.

Environmental factors also have a significant relationship to a household's probability of being poor. In particular, households living close to roads, in large counties and with irrigation systems have a low probability of being poor.

TABLE 5. **Coefficient estimates, standard error, *z*-value of the autologistic model**

Coefficients	Estimate	Std. error	*z* value	Pr(>|z|)	Significance
Percentage adults illiterate in household	2.69E-02	1.02E-01	0.263	0.792368	
Percentage persons with diploma	-2.12E+00	2.49E-01	-8.519	2.00E-16	***
Adequate home	-7.62E-02	6.33E-02	-1.204	0.228537	
Home with drinking-water	-6.85E-03	5.01E-02	-0.137	0.891297	
Home with adequate toilet	-1.07E-01	6.20E-02	-1.727	0.084099	.
Home with adequate wall	-1.83E-01	4.68E-02	-3.91	9.24E-05	***
Home with public electricity network	8.40E-02	7.28E-02	1.154	0.248359	
Waste collection by truck	-3.13E-01	5.24E-02	-5.977	2.28E-09	***
Persons per room	3.75E-01	1.69E-02	22.142	2.00E-16	***
Population	7.98E-06	2.99E-06	2.67	0.007594	**
Mortality rate (‰)	5.81E-03	2.48E-03	2.347	0.018943	*
Number of babies	-2.80E-04	1.39E-04	-2.011	0.044284	*
Slippery and landslide	4.06E-01	2.05E-01	1.987	0.046905	*
Sulifluxion	3.61E-01	1.81E-01	1.997	0.045802	*
Temperate dry	4.93E-01	2.22E-01	2.221	0.026349	*
Temperate humid	2.88E-01	2.11E-01	1.363	0.173001	
Hot and temperate	1.88E-01	2.22E-01	0.846	0.397364	
Hot and temperate humid	8.46E-01	3.73E-01	2.264	0.023572	*
Flooding area	1.77E-01	1.48E-01	1.196	0.231526	
Volcano area	6.42E-02	1.50E-01	0.428	0.668536	
Spatial correlation variable (y^*)	-1.35E-03	3.51E-04	-3.851	0.000118	***
Rural or urban	5.74E-02	6.01E-02	0.955	0.339784	
People < 5 km from road	-2.34E-06	8.70E-07	-2.692	0.007101	**
People 5-15 km from road	1.95E-06	3.00E-06	0.651	0.515001	
People > 15 km from road	-8.83E-07	1.20E-05	-0.074	0.941289	
County surface (km^2)	-3.15E-05	7.60E-06	-4.142	3.45E-05	***
Cereal production coefficient	3.83E-04	2.05E-04	1.871	0.061304	.
Protected area	1.01E-01	7.59E-02	1.332	0.18282	
> 35% of irrigation area	-2.18E-01	7.44E-02	-2.925	0.00345	**
Closed forest	3.06E-02	6.90E-02	0.443	0.657655	
Arable land (30-60%)	3.01E-02	7.61E-02	0.395	0.692683	
Arable land (> 60%)	3.35E-01	2.15E-01	1.556	0.119686	

Significance codes: 0 '****' 0.001 '***' 0.01 '**' 0.05 '.' 0.1 ' ' 1.

Finally, it is important to underline the significant effect of the spatial correlation variable (y^*) that denotes the presence of clusters in the spatial distribution of poverty and the influence between neighbour households on the probability of being poor.

The estimated parameters of the autologistic model (Equation 12) were applied to the data in the county database (INFOPLAN) in order to predict the distribution of poverty across all the counties in Ecuador. The percentage of poor households in each county was obtained from Equation 15.

In order to count the number of poor individuals, the average households' components in each county were multiplied by the number of poor families in each county (Figure 1).

Table 6. **Poverty level comparison (in ascending order): econometric estimation and ECV data**

County	% poor persons	County	Probit model (%)	County	Spatial probit model (%)
Milagro	16.7	Quito	7.6	Quito	10.8
Rumiñahui	17.0	Guayaquil	21.5	Rumiñahui	23.2
Quito	29.2	Rumiñahui	21.7	Cuenca	29.5
Zamora	31.5	Cuenca	29.6	Guayaquil	31.7
Quevedo	32.1	Machala	31.4	Machala	32.0
S. Dom. Colorados	32.7	Portoviejo	33.2	Portoviejo	33.5
Tena	33.3	Ambato	34.6	Ambato	35.2
Guayaquil	33.9	Manta	35.4	S. Dom. Colorados	36.5
Santa Rosa	34.0	S. Dom. Colorados	36.8	Manta	37.5
Machala	35.8	Quevedo	40.6	Quevedo	42.8
Pastaza	36.9	Esmeraldas	42.6	Pastaza	44.8
Ambato	38.2	Pastaza	43.6	Baños	45.6
Manta	40.7	Riobamba	44.6	Riobamba	45.8
Cuenca	42.3	Baños	45.3	Esmeraldas	45.8
Portoviejo	43.1	Loja	46.5	Loja	47.5
Baños	45.5	Milagro	47.8	Milagro	49.4
Playas	45.5	Santa Rosa	51.0	Santa Rosa	52.2
Morona	47.1	Cayambe	53.2	Cayambe	52.3
Cayambe	48.1	Piñas	53.5	Piñas	53.7
Santa Elena	49.6	Sucre	54.2	Sucre	55.9
S. Miguel De Los Bancos	50.0	Ibarra	56.8	Zamora	57.7
Gualaquiza	50.8	Zamora	57.7	Ibarra	58.9
Sucua	50.9	Playas	58.0	Playas	60.0
Esmeraldas	51.7	Tulcan	59.6	Urdaneta	60.1
Ventanas	51.9	Urdaneta	60.6	Junin	60.9
Urdaneta	52.1	Junin	61.3	Tulcan	61.1
Montecristi	54.5	Chimbo	62.2	Chimbo	62.2
Ibarra	54.7	Montecristi	64.1	Sucua	64.0
Quero	56.9	Morona	64.2	S. Miguel De Los Bancos	64.2
Sucre	57.6	S. Miguel De Los Bancos	64.4	Gualaquiza	64.2
Tulcan	58.8	El Empalme	65.5	Morona	64.3
Santa Lucia	59.3	Sucua	65.8	Montecristi	66.8
Chimbo	60.0	Gualaquiza	65.9	Tena	67.1
Loja	60.4	Tena	67.2	El Empalme	67.2
Junin	61.1	La Troncal	68.0	Quero	67.6
Lago Agrio	61.1	Ventanas	68.3	La Troncal	68.9
Catamayo	63.6	Quero	69.3	Ventanas	70.0
Riobamba	64.3	Catamayo	70.3	San Juan Bosco	71.3
Piñas	65.3	Jipijapa	70.8	Jipijapa	71.5
San Juan Bosco	67.8	Santa Elena	70.9	Catamayo	72.2
Santa Isabel	69.1	Calvas	71.4	Calvas	72.7
La Troncal	71.2	San Juan Bosco	71.9	Santa Elena	72.9
El Empalme	72.2	Montufar	72.3	Lago Agrio	73.1

(Continued)

Table 6. (Continued)

County	% poor persons	County	Probit model (%)	County	Spatial probit model (%)
Saraguro	72.4	Lago Agrio	72.7	Montufar	73.6
Calvas	72.5	Gualaceo	74.9	Gualaceo	75.8
Montufar	73.4	Zapotillo	76.5	Zapotillo	77.5
Quininde	76.4	Quininde	79.5	Quininde	78.6
Cotacachi	80.6	Santa Lucia	80.5	Santa Lucia	80.2
Zapotillo	80.9	Cotacachi	80.9	Cotacachi	80.8
Jipijapa	81.9	Santa Isabel	83.0	Santa Isabel	82.6
Gualaceo	82.2	Saraguro	85.6	Saraguro	85.7
Urbina Jado	84.1	Urbina Jado	87.8	Urbina Jado	88.8
Guamote	87.5	Guamote	95.9	Guamote	96.3
Total	**47.7**		**45.6**		**48.2**

Notes: Spearman coefficients: ECV and probit model = 0.781; ECV and spatial probit model = 0.791; probit model and spatial probit model = 0.995.

The first validation test draws its conclusion about the reliability of the incidence of poverty from the counties' ranking. The higher the coefficient of rank correlation for the counties, the higher the likelihood that the ranking of counties established by these estimates will also be highly correlated with the ranking established by the ECV data. Table 6 compares the results using the probit model and the spatial probit model with the estimates obtained directly from the ECV data.

Figure 2 shows the aggregate poverty situation at the province level. Table 7 compares these results with those obtained by applying a probit regression without the spatial component.

Figure 1. **Percentage of poor people in each county**

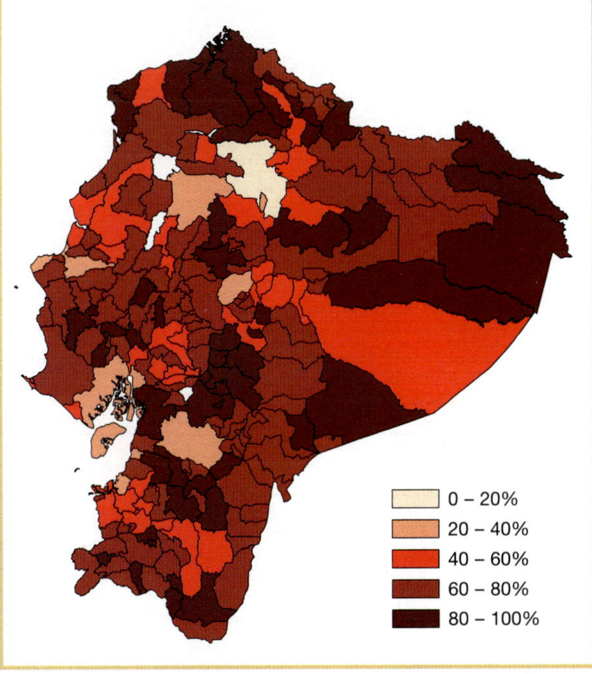

Figure 2. **Percentage of poor people in each province**

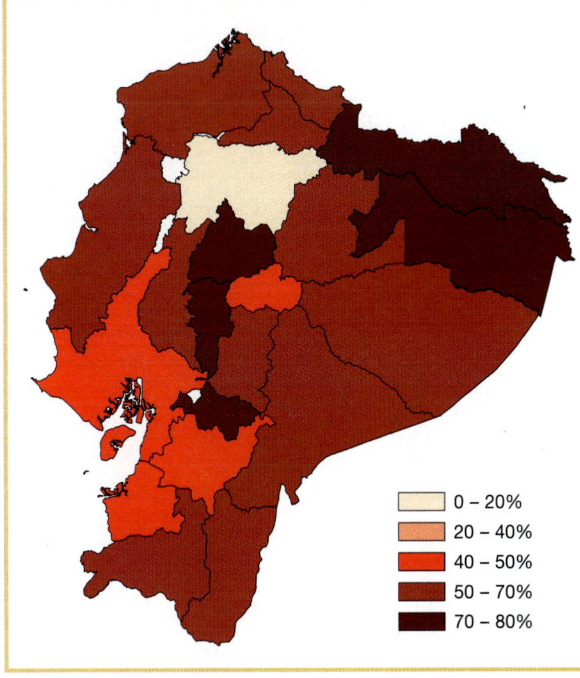

Table 7. **Comparison between probit and spatial probit in the headcount index and in the percentage of poor people in the provinces of Ecuador, total sample**

Province	Total poor Bigman (a)	Total poor spatial p. (b)	Total pop.	% poor Bigman (a)	% poor spatial p. (b)	Differences [((b-a)/b)*100]
Azuay	235 579	235 692	506 090	46.5	46.6	0.048
Bolivar	107 848	108 660	155 088	69.5	70.1	0.747
Ca_ar	138 804	140 292	185 927	74.7	75.5	1.061
Carchi	97 513	98 698	141 482	68.9	69.8	1.200
Chimborazo	242 872	245 923	362 430	67.0	67.9	1.240
Cotopaxi	196 531	197 070	276 324	71.1	71.3	0.273
El Oro	199 160	201 868	412 725	48.3	48.9	1.342
Esmeraldas	205 404	209 311	315 449	65.1	66.4	1.867
Guayas	941 136	1 115 349	2517 398	37.4	44.3	15.620
Imbabura	178 447	181 566	265 499	67.2	68.4	1.718
Loja	254 383	257 534	384 545	66.2	67.0	1.223
Los Rios	324 256	330 755	527 559	61.5	62.7	1.965
Manabi	570 799	576 256	1 031 927	55.3	55.8	0.947
Morona Santiago	58 653	57 962	84 216	69.6	68.8	-1.193
Napo	40 434	40 063	57 316	70.5	69.9	-0.926
Orellana	34 845	34 930	46 328	75.2	75.4	0.243
Pastaza	21 227	21 448	41 554	51.1	51.6	1.029
Pichincha	264 323	306 122	1 756 228	15.1	17.4	13.654
Sucumbios	58 287	57 895	76 952	75.7	75.2	-0.677
Tungurahua	157 685	157 312	361 980	43.6	43.5	-0.237
Zamora Chinchipe	44 769	44 517	66 167	67.7	67.3	-0.567
Total	**4 372 957**	**4 619 221**	**9 573 184**	**45.6**	**48.2**	**5.394**

Table 8. **Comparison between probit and spatial probit in the headcount index and in the percentage of poor people in the provinces of Ecuador, rural sample**

Province	Total poor Bigman (a)	Total poor spatial p. (b)	Total pop.	% poor Bigman (a)	% poor spatial p. (b)	Differences [((b-a)/b)*100]
Azuay	207 797	204 639	302 537	68.7	67.6	-1.543
Bolivar	103 282	103 631	139 358	74.1	74.4	0.337
Ca_ar	110 052	110 075	130 958	84.0	84.1	0.021
Carchi	75 624	75 176	92 440	81.8	81.3	-0.596
Chimborazo	217 438	217 950	255 942	85.0	85.2	0.235
Cotopaxi	176 080	175 486	214 655	82.0	81.8	-0.338
El Oro	89 114	87 217	123 728	72.0	70.5	-2.175
Esmeraldas	144 262	143 221	172 741	83.5	82.9	-0.727
Guayas	309 035	307 138	414 686	74.5	74.1	-0.618
Imbabura	123 950	123 367	142 879	86.8	86.3	-0.472
Loja	208 061	207 866	250 089	83.2	83.1	-0.094
Los Rios	227 863	227 714	294 834	77.3	77.2	-0.065
Manabi	391 569	386 384	542 961	72.1	71.2	-1.342
Morona Santiago	55 657	54 802	75 970	73.3	72.1	-1.560
Napo	38 381	37 838	49 443	77.6	76.5	-1.437
Orellana	31 294	31 148	38 523	81.2	80.9	-0.468
Pastaza	18 707	18 600	27 116	69.0	68.6	-0.575
Pichincha	119 782	112 367	367 646	32.6	30.6	-6.599
Sucumbios	50 358	49 597	63 787	78.9	77.8	-1.535
Tungurahua	134 471	131 623	217 213	61.9	60.6	-2.164
Zamora Chinchipe	42 364	41 937	58 119	72.9	72.2	-1.018
Total	**2 875 141**	**2 847 776**	**3 975 625**	**72.3**	**71.6**	**-0.977**

Table 9. **Comparison between probit and spatial probit in the headcount index and in the percentage of poor people in the provinces of Ecuador, urban sample**

Province	Total poor Bigman (a)	Total poor spatial p. (b)	Total pop.	% poor Bigman (a)	% poor spatial p. (b)	Differences [((b-a)/b)*100]
Azuay	27 783	31 053	203 553	13.6	27.0	10.531
Bolivar	4 566	5 029	15 730	29.0	25.7	9.198
Ca_ar	28 752	30 216	54 969	52.3	52.6	4.847
Carchi	21 889	23 522	49 042	44.6	42.8	6.942
Chimborazo	25 434	27 973	106 488	23.9	28.3	9.075
Cotopaxi	20 452	21 585	61 669	33.2	36.3	5.249
El Oro	110 046	114 651	288 997	38.1	43.7	4.017
Esmeraldas	61 141	66 090	142 708	42.8	48.4	7.488
Guayas	632 101	808 211	2 102 712	30.1	39.6	21.790
Imbabura	54 497	58 199	122 620	44.4	46.3	6.361
Loja	46 323	49 667	134 456	34.5	33.9	6.734
Los Rios	96 394	103 040	232 725	41.4	46.3	6.451
Manabi	179 230	189 872	488 966	36.7	40.7	5.605
Morona Santiago	2 996	3 160	8 246	36.3	36.9	5.169
Napo	2 052	2 225	7 873	26.1	31.5	7.774
Orellana	3 551	3 781	7 805	45.5	49.4	6.098
Pastaza	2 520	2 848	14 438	17.5	31.7	11.505
Pichincha	144 542	193 756	1 388 582	10.4	19.6	25.400
Sucumbios	7 929	8 298	13 165	60.2	62.9	4.453
Tungurahua	23 214	25 689	144 767	16.0	23.2	9.636
Zamora Chinchipe	2 405	2 579	8 048	29.9	28.9	6.776
Total	**1 497 815**	**1 771 445**	**5 597 559**	**26.7**	**31.6**	**15.506**

With the spatial probit regression, the percentage of poor persons in the total population is 48 percent, compared with 45 percent for the probit regression. The difference (5.39 percent) is very significant. For the rural population, the corresponding figures for the spatial probit regression and the probit regression are 71 and 72 percent respectively (Table 8). The differences among the urban population are more evident (Table 9), even if the spatial autocorrelation is very small. These

TABLE 10. **Comparison of three methods for estimating poverty, percentage of poor people in Ecuador**

Area	Method		
	Lanjouw *et al.* (1995)	Bigman *et al.*	Spatial probit
Costa	**54**	**46**	**50**
Urban	43	33	39
Rural	75	75	74
Sierra	**58**	**42**	**44**
Urban	42	17	20
Rural	78	69	69
Oriente	**65**	**69**	**69**
Urban	47	36	39
Rural	70	75	75
Ecuador	**56**	**45**	**48**
Urban	42	26	31
Rural	76	72	71

Figure 3. **Comparison between the Lanjouw *et al.* method and spatial probit at regional level**

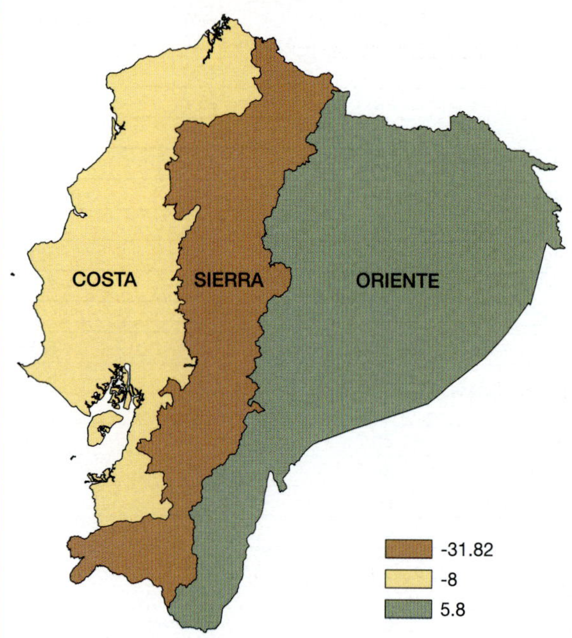

Figure 4. **Comparison between the Bigman *et al.* method and spatial probit at regional level**

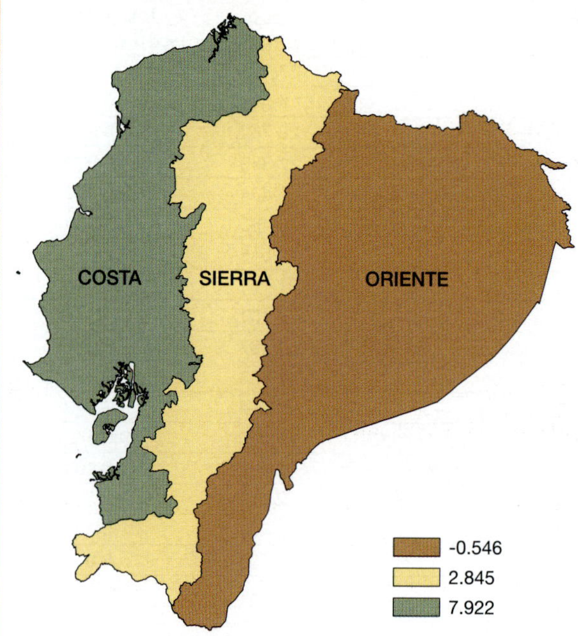

differences are very important if poverty distribution is analysed at the county level (Table A.1 in Annex A).

The method applied in this study for poverty mapping at the level of small geographical areas provides a more accurate specification of poverty by incorporating a spatial component into the classical model regression.

Recalling the comparison performed in Chapter 3, Table 10 and Figures 3 and 4 compare the percentage of poor individuals at the regional level as obtained with spatial probit regression with the percentage of poor persons using the Lanjouw *et al.* and Bigman *et al.* methods.

Chapter 6. **A simple simulation**

In order to evaluate the effectiveness of community targeting, a simple simulation was carried out.

The simulation study was defined by three interventions aimed at reducing poverty (the suggestions are plausible but not real):

- To increase the number of individuals with a diploma by 10 percent in each county.
- To decrease land erosion by 15 percent with respect to the total.
- To improve the public drinking-water network (to increase the number of homes with potable water by 10 percent in each county).

Figures 5 and 6 and Table 11 show the results and comparison between the percentage of poor individuals before and after the interventions.

The analysis of the simulation experiment highlights three important results:

- a general reduction of poverty in Ecuador (-7.375 percent);
- the largest decrease is in the counties where it was decided to reduce land erosion;

Table 11. **Comparison of the percentages of poor people before and after the simulation**

County	Before (a)	After (b)	Differences [((b-a)/b)*100]
Atahualpa	59.0	42.6	-38.686
Balao	77.8	65.4	-18.809
Cuenca	29.5	18.4	-60.590
Cuyabeno	83.9	73.4	-14.237
El Pan	71.9	56.7	-26.683
Eloy Alfaro	87.6	78.3	-11.871
Guayaquil	31.7	19.0	-66.628
Jama	73.5	60.3	-21.889
Logroño	85.4	73.9	-15.495
Mera	55.7	40.3	-38.122
Montufar	73.6	60.7	-21.277
Morona	64.3	50.5	-27.394
Nangaritza	72.1	58.4	-23.389
Olmedo	85.7	74.3	-15.271
Olmedo	72.4	72.2	-0.374
Pastaza	44.8	33.1	-35.255
Putumayo	86.0	76.0	-13.068
Quito	10.8	09.8	-10.481
Salcedo	70.2	56.3	-24.664
San Juan Bosco	71.3	55.9	-27.521
Sucumbios	84.0	73.3	-14.493
Urbina Jado	88.8	80.5	-10.328
Total	**48.2**	**44.8**	**-7.375**

Figure 5. **Percentage of poor people after the simulation experiment in each county**

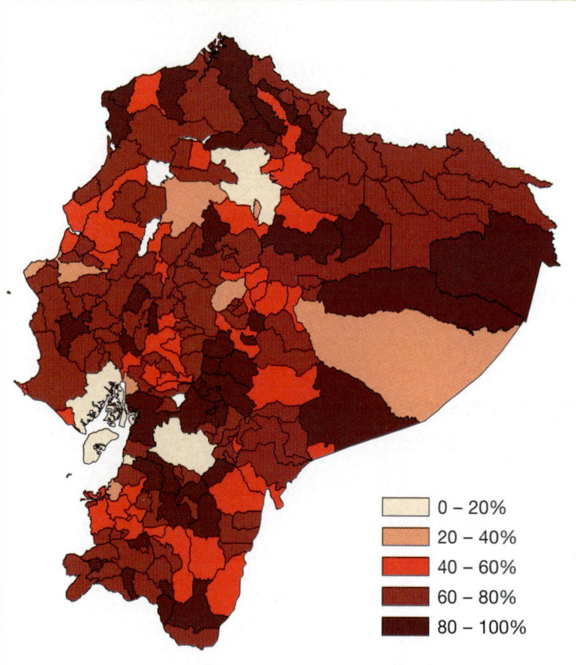

Figure 6. **Differences between before and after the simulation experiment in each county**

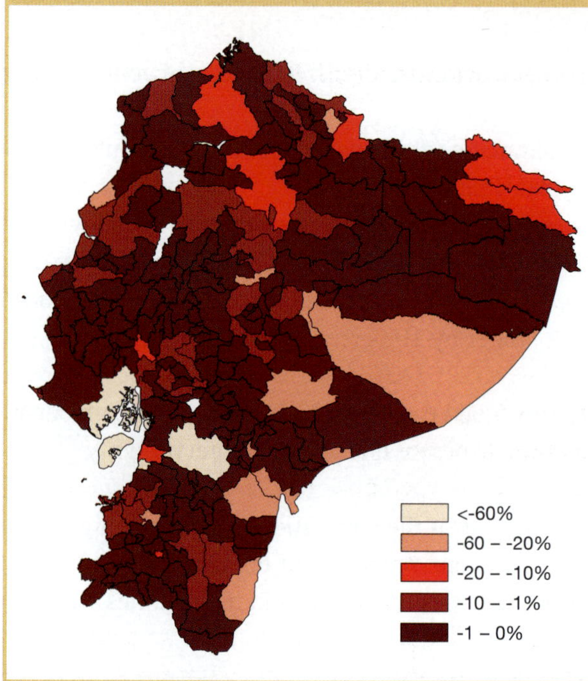

- increasing the number of individuals with a diploma in each county has a small effect on decreasing poverty.

Mapping poverty requires a clear definition of poverty and an index of poverty that is based on widely available data. The measures of poverty thus estimated must also have a known and well-defined degree of precision so that statistical statements can be made about the precision of the poverty measures in different areas. In order to address this problem, an experiment was carried out in which the definition of the poverty line was changed. Annex B reports the results of this experiment.

Chapter 7. **Conclusions**

There are some important implications of applying spatial statistical analysis to poverty-mapping studies.

First is the importance of taking the spatial dimension of the data into account. In fact, after having found a significant spatial correlation between the units, ignoring the spatial component in the regression analysis could lead to misleading estimates of the parameters. This may result in a large proportion of poor households being excluded from the allocation of transfers while a number of non-poor households might be declared potential beneficiaries.

Moreover, the use of spatial models in combination with the visual nature of the poverty maps, obtained from applying the spatial regression methods, may highlight unexpected relationships that would escape notice in a standard regression analysis.

One of the most important difficulties encountered in spatial analysis concerns the availability of skills and data. It would be useful to develop poverty maps by conducting the statistical analysis at a degree of disaggregation below broad regions; otherwise it is assumed implicitly that, within a region, the model of consumption is the same for all households irrespective of which province, county or community they reside in. However, all the data from the household survey (including measures of consumption) are usually based on samples. They are rarely sufficiently representative at low levels of aggregation to yield reliable estimates. At the same time, the census data, which are of sufficient size to allow disaggregation, have no information on consumption. Moreover, even where some subnational poverty data are routinely available, they are often not georeferenced. For this reason, the adoption of the small area estimation could offer a solution: a statistical technique that combines survey and census data to estimate poverty indicators for disaggregated geographic units (counties in this study). The only problem with this statistical tool is that many countries do not provide census and survey data simultaneously. Hence, when the temporal gap between these two kinds of data is large, census and survey information become incompatible.

In the context of data availability, other kinds of useful information about factors that influence poverty are not readily available for all countries, e.g. data on land use and the environment (e.g. for environmentally fragile ecological zones with no productive agricultural lands) and on accessibility. Accessibility concerns issues such as: access to health, water, education; facility location and infrastructure; transport, travel time and the road conditions of a zone. The availability of these kinds of data varies widely from country to country, and the design and selection of case studies must take this aspect into account.

The results of the fitted spatial model demonstrate the statistical significance of environmental variables. This suggests the presence of a poverty-environment relationship and hence the impact of environmental factors on the lives of the poor and on poverty reduction efforts. For this reason, environmental indicators could be an important tool for designing and evaluating poverty reduction strategies and they should be introduced into the statistical analysis.

The results obtained by the study have demonstrated the usefulness of the spatial analysis in poverty targeting compared with the classical statistical analysis. However, the significant data requirements of the spatial technique constitute a drawback compared with the simple regression analysis. The creation of an ad hoc database (georeferenced and geographically disaggregated) is time consuming. This problem could be overcome if all countries provided a unique database with socio-economic and demographic information, with data on food insecurity, geophysical parameters, etc. being as spatially disaggregated as possible and georeferenced.

Another problem of all polygon-based spatial analysis is the modifiable areal unit problem (MAUP): the areal units (administrative or political boundaries, agro-ecological zones, etc.) are arbitrary groupings and the data within each can be aggregated in an infinite number of ways (Nelson, 2001; Bigman and Deichmann, 2000). The implication is that different kinds of aggregation can lead to different results in the spatial analysis so that variables, parameters and processes that are important at one scale or unit are frequently not important or predictive at another scale or unit.

A definitive solution to minimize this effect remains to be found. Depending on the issue being examined, some methods have been used to solve the MAUP, for example: identifying the basic units and deriving optimal scales and zonal configurations (Openshaw and Taylor, 1981); conducting a sensitivity analysis (O'Neil *et al.*, 1988); and adopting tools such as convolution filtering, different methods of zonification including extending the concept of areal units to measures such as time, accessibility, etc. (Nelson, 2001).

Concerning the concept and measurement of poverty, the analysis in this study adopted an index calculated using both consumption and the poverty line. Other kinds of index could be used as multiple indicators exist for poverty (Annex B). Poverty can be evaluated using: economic measures such as monetary indicators of households well-being (expenditures, income, consumption, etc.); demographic indicators (gender and age of head of the household, household size, infant mortality rates); and environmental and health measures (access to safe water and sanitation, time spent by household to collect water, cereal production for a family, prevalence of acute infections, disability adjusted life years) (Shyamsundar, 2002).

The choice of one indicator rather than another usually depends on the availability of the data, and on the practical implications in terms of time, costs and technical requirements for constructing the index.

The consequence of using a particular index is that different indicators can lead to different results of the analysis, and so to alternative poverty rankings.

One solution, albeit time consuming, could be to apply different kinds of indicators to the same analysis and then to compare and evaluate the implications of each index. In the application on the Ecuador data, the index adopted was the same used in other studies in order to enable the results of this study to be compared with results obtained using other common methods.

References

Anselin, L. 1992. *Spatial econometrics: method and models.* Boston, USA, Kluwer Academic Publishers.

Anselin, L. 1995. Local indicator of spatial association - LISA. *Geog. Anal.*, 27: 93–116.

Bao, S. & Henry, M.S. 1996. Heterogeneity issues in local measurements of spatial association. *Geog. Sys.*, 3: 1–13.

Bao, S. 1999. An overview of spatial statistics. University of Michigan, USA, China Data Center.

Besag, J.E. 1975. Statistical analysis of non-lattice data. *Statistician*, 24: 179–195.

Besag, J.E. 1977. Efficiency of pseudo-likelihood estimation for simple Gaussian fields. *Biometrika*, 64: 616–618.

Bigman, D. & Deichmann, U. 2000. Geographic targeting for poverty alleviation. In D. Bigman & H. Fofack, eds. *Geographic targeting for poverty alleviation.* Methodology and application. World Bank Regional and Sectoral Studies.

Bigman, D., Dercon, S., Guillaume, D. & Lambotte, M. 2000. Community targeting for poverty reduction in Burkina Faso. *World Bank Econ. Rev.*, 14, No. 1: 167–193.

Cliff, A.D. & Ord, J.K. 1981. *Spatial processes. Models and applications.* London, Pion Limited.

Deichmann, U. 1999. Geographic aspects of inequality and poverty. (available at: http://www.worldbank.org/poverty/inequal/index.htm).

Elbers, C., Lanjouw, J.O. & Lanjouw, P. 2001. Welfare in villages and towns: micro-level estimation of poverty and inequality. World Bank.

Getis, A., Getis, O. & Keith, J. 1992. The analysis of spatial association by the use of distance statistics. *Geog. Anal.*, 24: 189–206.

Hentschel, J., Lanjouw, J.O., Lanjouw, P. & Poggi, J. 2000. Combining census and survey data to trace the spatial dimensions of poverty: a case study of Ecuador. *World Bank Econ. Rev.*, 14, No. 1: 147–165.

Hentschel, J. & Lanjouw, P. 1996. *Constructing an indicator of consumption for the analysis of poverty.* World Bank, LSMS Working Paper No. 124.

Jalan, J. & Ravallion, M. 1998. *Geographic poverty traps.* Word Bank, May 15. pp. 1–31.

Lanjouw, P. 1998. *Ecuador's rural nonfarm sector as a route out of poverty.* World Bank, Policy Research Working Paper No. 1904.

Moran, P.A.P. 1950. Notes on continuous stochastic phenomena. *Biometrika*, 37: 17.

Nelson, A. 2001. Analyzing data across geographic scales in Honduras: detecting levels of organization within systems. *Ag. Ecosys. Env.*

O'Neil, R.V., Krummel, J.R., Gardner, R.H., Sugihara, G., Jackson, B., De Angelis, D., Milne, B.T., Turner, M.G., Zygmunt, B., Christensen, S.V., Dale, V.H. & Graham, R.L. 1988. Indices of landscape pattern. *Land. Eco.*, 1: 153–162.

Openshaw, S. & Taylor, P. 1981. The modifiable unit problem. *In* N. Wrigley, ed. *Quantitative geography, a British view*. London, Pion. pp. 127–144.

Shyamsundar, P. 2002. *Poverty environment indicators.* World Bank Environment Department, Paper No. 84.

Stauss, D. & Ikeda, M. 1990. Pseudo-likelihood estimation for social networks. *J. Am. Stat. Ass.*, 85: 204–212.

Annex A. **Tables and maps**

Table A.1. **Comparison between probit and spatial probit in the headcount index and in the percentage of poor people at county level, total sample**

County	Total poor		Total pop.	% poor		Difference [((b-a)/b)*100]
	Bigman (a)	Spatial p. (b)		Bigman (a)	Spatial p. (b)	
Cuenca	98 079	97 735	331 028	29.6	29.5	-0.351
Giron	10 141	10 088	13 191	76.9	76.5	-0.525
Gualaceo	26 664	26 969	35 586	74.9	75.8	1.130
Nabon	11 716	11 768	14 606	80.2	80.6	0.436
Paute	16 207	16 327	21 887	74.0	74.6	0.736
Pucara	14 586	14 626	16 626	87.7	88.0	0.272
San Fernando	3 324	3 342	4 298	77.3	77.8	0.526
Santa Isabel	14 242	14 181	17 160	83.0	82.6	-0.434
Sigsig	20 566	20 619	25 103	81.9	82.1	0.258
Oña	2 461	2 473	3 244	75.9	76.2	0.503
Chordeleg	7 475	7 467	9 683	77.2	77.1	-0.111
El Pan	2 672	2 653	3 690	72.4	71.9	-0.750
Sevilla De Oro	4 928	4 902	6 553	75.2	74.8	-0.529
Guachapala	2 518	2 543	3 435	73.3	74.0	1.001
Guaranda	48 353	49 084	66 241	73.0	74.1	1.491
Chillanes	15 785	15 954	20 477	77.1	77.9	1.060
Chimbo	9 699	9 700	15 602	62.2	62.2	0.005
Echeandia	6 092	6 105	9 821	62.0	62.2	0.216
San Miguel	18 048	18 004	28 044	64.4	64.2	-0.242
Caluma	6 095	5 998	9 828	62.0	61.0	-1.603
Las Naves	3 777	3 814	5 075	74.4	75.2	0.974
Azogues	41 848	42 513	61 570	68.0	69.0	1.565
Biblian	16 665	16 719	22 286	74.8	75.0	0.323
Cañar	46 606	47 010	55 259	84.3	85.1	0.860
La Troncal	22 138	22 412	32 540	68.0	68.9	1.224
El Tambo	6 116	6 152	6 974	87.7	88.2	0.582
Deleg	5 431	5 485	7 298	74.4	75.2	0.991
Tulcan	37 614	38 598	63 158	59.6	61.1	2.550
Bolivar (De Carchi)	12 471	12 442	15 157	82.3	82.1	-0.235
Espejo	9 541	9 469	13 188	72.3	71.8	-0.767
Mira	11 484	11 446	14 040	81.8	81.5	-0.331
Montufar	21 294	21 686	29 454	72.3	73.6	1.806
San Pedro De Huaca	5 109	5 058	6 485	78.8	78.0	-1.020
Latacunga	75 061	75 130	117 603	63.8	63.9	0.092
La Mana	14 433	14 564	20 960	68.9	69.5	0.897
Pangua	13 058	13 149	16 814	77.7	78.2	0.688
Pujili	35 343	35 554	44 343	79.7	80.2	0.592
Salcedo	31 799	31 808	45 322	70.2	70.2	0.027

(Continued)

Table A.1. (Continued)

County	Total poor		Total pop.	% poor		Difference [((b-a)/b)*100]
	Bigman (a)	Spatial p. (b)		Bigman (a)	Spatial p. (b)	
Saquisili	9 893	9 857	12 829	77.1	76.8	-0.375
Sigchos	16 943	17 010	18 453	91.8	92.2	0.389
Riobamba	73 080	74 995	163 779	44.6	45.8	2.553
Alausi	34 109	34 473	39 309	86.8	87.7	1.055
Colta	41 615	42 052	47 568	87.5	88.4	1.038
Chambo	7 636	7 656	9 463	80.7	80.9	0.268
Chunchi	11 660	11 646	13 490	86.4	86.3	-0.119
Guamote	26 917	27 013	28 058	95.9	96.3	0.355
Guano	28 493	28 653	37 106	76.8	77.2	0.561
Pallatanga	7 864	7 919	9 482	82.9	83.5	0.699
Penipe	5 788	5 730	7 078	81.8	81.0	-1.005
Cumanda	5 711	5 785	7 097	80.5	81.5	1.282
Machala	49 491	50 447	157 607	31.4	32.0	1.895
Arenillas	9 702	9 809	18 314	53.0	53.6	1.092
Atahualpa	3 728	3 626	6 142	60.7	59.0	-2.803
Balsas	2 651	2 587	4 066	65.2	63.6	-2.488
Chilla	2 055	2 033	2 767	74.3	73.5	-1.083
El Guabo	19 871	19 817	28 058	70.8	70.6	-0.271
Huaquillas	17 487	17 985	27 806	62.9	64.7	2.770
Marcabeli	3 281	3 254	4 889	67.1	66.6	-0.836
Pasaje	28 665	29 537	51 406	55.8	57.5	2.952
Piñas	11 685	11 723	21 843	53.5	53.7	0.322
Portovelo	5 811	5 838	10 257	56.7	56.9	0.458
Santa Rosa	25 926	26 525	50 852	51.0	52.2	2.259
Zaruma	14 864	14 803	23 852	62.3	62.1	-0.410
Las Lajas	3 943	3 883	4 866	81.0	79.8	-1.529
Esmeraldas	57 564	61 927	135 145	42.6	45.8	7.044
Eloy Alfaro	22 267	22 237	25 389	87.7	87.6	-0.135
Muisne	18 278	18 095	22 537	81.1	80.3	-1.007
Quininde	57 135	56 509	71 901	79.5	78.6	-1.108
San Lorenzo	18 043	18 352	22 552	80.0	81.4	1.686
Atacames	15 801	15 980	18 014	87.7	88.7	1.122
Rio Verde	16 317	16 211	19 911	81.9	81.4	-0.649
Guayaquil	338 200	498 000	1 570 011	21.5	31.7	32.088
Alfredo Baquerizo Moreno	12 233	12 611	16 253	75.3	77.6	2.999
Balao	9 569	9 730	12 514	76.5	77.8	1.656
Balzar	30 523	31 810	43 965	69.4	72.4	4.044
Colimes	15 342	15 190	19 232	79.8	79.0	-1.001
Daule	44 471	44 988	65 534	67.9	68.6	1.150
Duran	31 164	34 078	85 196	36.6	40.0	8.551
El Empalme	38 198	39 192	58 351	65.5	67.2	2.538
El Triunfo	15 875	16 065	24 551	64.7	65.4	1.186
Milagro	55 693	57 488	116 397	47.8	49.4	3.121

(Continued)

Table A.1. (Continued)

County	Total poor		Total pop.	% poor		Difference [((b-a)/b)*100]
	Bigman (a)	Spatial p. (b)		Bigman (a)	Spatial p. (b)	
Naranjal	38 222	37 587	39 466	96.8	95.2	-1.689
Naranjito	13 457	13 806	25 608	52.6	53.9	2.524
Palestina	7 669	7 830	11 191	68.5	70.0	2.055
Pedro Carbo	22 128	22 803	31 628	70.0	72.1	2.959
Salinas	16 511	17 042	32 434	50.9	52.5	3.111
Samborondon	18 948	19 629	33 965	55.8	57.8	3.468
Santa Elena	59 544	61 205	84 010	70.9	72.9	2.714
Santa Lucia	22 006	21 922	27 342	80.5	80.2	-0.379
Urbina Jado	38 462	38 904	43 812	87.8	88.8	1.135
Yaguachi	26 569	27 092	39 323	67.6	68.9	1.929
Playas	12 469	12 902	21 490	58.0	60.0	3.358
Simon Bolivar	13 632	13 591	17 240	79.1	78.8	-0.303
Coronel Marcelino Maridueña	5 424	5 450	11 078	49.0	49.2	0.474
Lomas De Sargentillo	7 353	7 579	10 410	70.6	72.8	2.992
Nobol (Piedrahita)	7 928	7 793	10 524	75.3	74.1	-1.732
La Libertad	31 389	32 763	53 108	59.1	61.7	4.193
General Antonio Elizalde (Bucay)	3 452	3 654	6 968	49.5	52.4	5.536
Isidro Ayora	4 705	4 646	5 797	81.2	80.1	-1.265
Ibarra	67 930	70 333	119 493	56.8	58.9	3.417
Antonio Ante	18 159	18 803	27 375	66.3	68.7	3.424
Cotacachi	26 914	26 855	33 250	80.9	80.8	-0.222
Otavalo	40 145	40 495	56 286	71.3	71.9	0.862
Pimampiro	14 035	14 020	15 359	91.4	91.3	-0.111
San Miguel De Urcuqui	11 263	11 061	13 736	82.0	80.5	-1.825
Loja	67 180	68 580	144 493	46.5	47.5	2.041
Calvas	20 998	21 373	29 398	71.4	72.7	1.757
Catamayo	15 726	16 133	22 357	70.3	72.2	2.520
Celica	11 264	11 325	14 329	78.6	79.0	0.538
Chaguarpamba	8 111	8 179	9 850	82.3	83.0	0.838
Espindola	15 830	15 815	18 191	87.0	86.9	-0.095
Gonzanama	13 908	13 822	17 276	80.5	80.0	-0.625
Macara	11 698	12 271	18 281	64.0	67.1	4.670
Paltas	21 991	22 123	28 605	76.9	77.3	0.597
Puyango	13 298	13 309	16 804	79.1	79.2	0.083
Saraguro	22 972	23 002	26 842	85.6	85.7	0.131
Sozoranga	8 338	8 330	9 729	85.7	85.6	-0.085
Zapotillo	7 830	7 931	10 234	76.5	77.5	1.269
Pindal	6 417	6 470	7 457	86.1	86.8	0.820
Quilanga	4 445	4 463	5 553	80.0	80.4	0.409
Olmedo	4 380	4 409	5 146	85.1	85.7	0.658
Babahoyo	57 078	57 445	105 471	54.1	54.5	0.638
Baba	22 580	22 500	29 406	76.8	76.5	-0.357
Montalvo	10 144	10 236	19 023	53.3	53.8	0.906

(Continued)

Table A.1. (Continued)

County	Total poor		Total pop.	% poor		Difference [((b-a)/b)*100]
	Bigman (a)	Spatial p. (b)		Bigman (a)	Spatial p. (b)	
Pueblo Viejo	19 168	19 364	22 662	84.6	85.4	1.013
Quevedo	42 248	44 541	104 175	40.6	42.8	5.147
Urdaneta	14 223	14 087	23 459	60.6	60.1	-0.963
Ventanas	39 955	40 950	58 494	68.3	70.0	2.428
Vinces	40 117	41 532	54 234	74.0	76.6	3.406
Palenque	15 977	16 215	18 242	87.6	88.9	1.471
Buena Fe	19 348	19 695	33 370	58.0	59.0	1.765
Valencia	19 472	19 683	27 270	71.4	72.2	1.072
Mocache	23 947	24 507	31 753	75.4	77.2	2.288
Portoviejo	67 054	67 537	201 861	33.2	33.5	0.715
Bolivar (De Manabi)	22 183	22 259	37 580	59.0	59.2	0.341
Chone	66 058	65 711	114 811	57.5	57.2	-0.529
El Carmen	33 673	34 089	54 070	62.3	63.0	1.220
Flavio Alfaro	16 385	16 244	23 613	69.4	68.8	-0.868
Jipijapa	48 970	49 453	69 177	70.8	71.5	0.976
Junin	11 177	11 102	18 243	61.3	60.9	-0.668
Manta	47 021	49 856	132 816	35.4	37.5	5.687
Montecristi	19 008	19 801	29 636	64.1	66.8	4.006
Pajan	34 159	34 311	42 446	80.5	80.8	0.444
Pichincha	21 760	21 474	28 827	75.5	74.5	-1.334
Rocafuerte	16 489	16 615	26 021	63.4	63.9	0.759
Santa Ana	32 976	32 592	49 164	67.1	66.3	-1.177
Sucre	35 410	36 532	65 322	54.2	55.9	3.071
Tosagua	20 935	20 891	31 778	65.9	65.7	-0.212
24 De Mayo	23 902	23 969	34 026	70.2	70.4	0.279
Pedernales	22 811	22 643	30 677	74.4	73.8	-0.739
Olmedo	7 402	7 365	10 166	72.8	72.4	-0.505
Puerto Lopez	10 599	10 782	13 630	77.8	79.1	1.698
Jama	7 404	7 376	10 039	73.8	73.5	-0.380
Jaramijo	5 421	5 651	8 024	67.6	70.4	4.077
Morona	13 194	13 219	20 562	64.2	64.3	0.191
Gualaquiza	8 254	8 039	12 518	65.9	64.2	-2.679
Limon	5 995	5 949	8 813	68.0	67.5	-0.770
Palanda	3 848	3 764	5 525	69.7	68.1	-2.232
Santiago	6 006	5 906	8 319	72.2	71.0	-1.685
Sucua	7 231	7 033	10 996	65.8	64.0	-2.817
Huamboya	2 976	2 949	3 922	75.9	75.2	-0.916
San Juan Bosco	2 336	2 316	3 249	71.9	71.3	-0.859
Taisha	5 664	5 673	6 666	85.0	85.1	0.146
Logroño	3 150	3 114	3 646	86.4	85.4	-1.143
Tena	22 397	22 365	33 337	67.2	67.1	-0.144
Archidona	11 194	11 027	13 650	82.0	80.8	-1.509
El Chaco	3 077	2 992	4 445	69.2	67.3	-2.863

(Continued)

Table A.1. (Continued)

County	Total poor		Total pop.	% poor		Difference [((b-a)/b)*100]
	Bigman (a)	Spatial p. (b)		Bigman (a)	Spatial p. (b)	
Quijos	2 329	2 251	3 847	60.5	58.5	-3.464
Carlos Julio Arosemena Tola	1 437	1 428	2 037	70.5	70.1	-0.598
Pastaza	12 989	13 336	29 780	43.6	44.8	2.603
Mera	3 427	3 310	5 947	57.6	55.7	-3.557
Santa Clara	1 631	1 620	2 233	73.0	72.5	-0.712
Arajuno	3 180	3 183	3 594	88.5	88.6	0.085
Quito	104 404	148 638	1 371 461	7.6	10.8	29.760
Cayambe	24 991	24 546	46 938	53.2	52.3	-1.810
Mejia	22 188	21 238	46 687	47.5	45.5	-4.474
Pedro Moncayo	7 377	7 047	15 718	46.9	44.8	-4.686
Rumiñahui	10 016	10 715	46 215	21.7	23.2	6.529
Sto Domingo De Los Colorados	70 145	69 634	190 825	36.8	36.5	-0.734
San Miguel De Los Bancos	10 449	10 442	16 928	61.7	61.7	-0.062
Pedro Vicente Maldonado	4 669	4 511	7 681	60.8	58.7	-3.494
Puerto Quito	10 086	9 351	13 775	73.2	67.9	-7.860
Ambato	78 805	80 133	227 790	34.6	35.2	1.657
Baños	6 978	7 027	15 416	45.3	45.6	0.693
Cevallos	3 452	3 317	5 964	57.9	55.6	-4.074
Mocha	4 329	4 212	6 368	68.0	66.2	-2.776
Patate	5 932	5 756	10 292	57.6	55.9	-3.072
Quero	11 086	10 812	15 997	69.3	67.6	-2.531
Pelileo	21 374	20 768	37 619	56.8	55.2	-2.918
Pillaro	19 129	18 835	33 369	57.3	56.4	-1.562
Tisaleo	6 599	6 452	9 165	72.0	70.4	-2.270
Zamora	12 552	12 547	21 743	57.7	57.7	-0.040
Chinchipe	5 243	5 218	6 989	75.0	74.7	-0.483
Nangaritza	3 137	3 111	4 316	72.7	72.1	-0.827
Yacuambi	2 751	2 766	3 448	79.8	80.2	0.542
Yantzaza	7 775	7 669	11 856	65.6	64.7	-1.385
El Pangui	4 295	4 274	6 054	70.9	70.6	-0.509
Centinela Del Condor	4 504	4 450	6 331	71.1	70.3	-1.199
Palanda	4 513	4 483	5 430	83.1	82.6	-0.669
Lago Agrio	28 186	28 314	38 753	72.7	73.1	0.451
Gonzalo Pizarro	3 290	3 206	4 472	73.6	71.7	-2.615
Putumayo	4 145	4 122	4 794	86.5	86.0	-0.548
Shushufindi	14 736	14 462	18 977	77.7	76.2	-1.899
Sucumbios	2 090	2 049	2 441	85.6	84.0	-2.004
Cascales	3 719	3 644	5 014	74.2	72.7	-2.067
Cuyabeno	2 121	2 098	2 501	84.8	83.9	-1.061
Orellana	12 626	12 836	19 174	65.8	66.9	1.639
Aguarico	2 733	2 738	3 150	86.8	86.9	0.196
La Joya De Los Sachas	12 571	12 458	16 193	77.6	76.9	-0.911
Loreto	6 915	6 897	7 811	88.5	88.3	-0.251
Total	**4 372 957**	**4 619 221**	**9 573 184**	**45.6**	**48.2**	**5.394**

Figure A.1. **Comparison between probit method and spatial probit at county level, total sample**

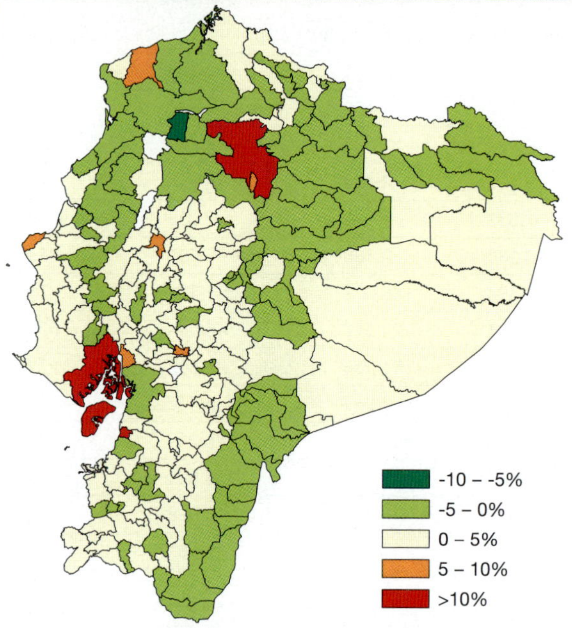

Table A.2. **Comparison between probit and spatial probit in the headcount index and in the percentage of poor people at county level, rural sample**

County	Total poor		Total pop.	% poor		Difference [((b-a)/b)*100]
	Bigman (a)	Spatial p. (b)		Bigman (a)	Spatial p. (b)	
Cuenca	74 407	71 075	136 047	54.7	52.2	-4.687
Giron	10 141	10 088	13 191	76.9	76.5	-0.525
Gualaceo	22 554	22 576	27 014	83.5	83.6	0.101
Nabon	11 716	11 768	14 606	80.2	80.6	0.436
Paute	16 207	16 327	21 887	74.0	74.6	0.736
Pucara	14 586	14 626	16 626	87.7	88.0	0.272
San Fernando	3 324	3 342	4 298	77.3	77.8	0.526
Santa Isabel	14 242	14 181	17 160	83.0	82.6	-0.434
Sigsig	20 566	20 619	25 103	81.9	82.1	0.258
Oña	2 461	2 473	3 244	75.9	76.2	0.503
Chordeleg	7 475	7 467	9 683	77.2	77.1	-0.111
El Pan	2 672	2 653	3 690	72.4	71.9	-0.750
Sevilla De Oro	4 928	4 902	6 553	75.2	74.8	-0.529
Guachapala	2 518	2 543	3 435	73.3	74.0	1.001
Guaranda	43 786	44 055	50 511	86.7	87.2	0.611
Chillanes	15 785	15 954	20 477	77.1	77.9	1.060
Chimbo	9 699	9 700	15 602	62.2	62.2	0.005
Echeandia	6 092	6 105	9 821	62.0	62.2	0.216
San Miguel	18 048	18 004	28 044	64.4	64.2	-0.242
Caluma	6 095	5 998	9 828	62.0	61.0	-1.603
Las Naves	3 777	3 814	5 075	74.4	75.2	0.974

(Continued)

Table A.2. (Continued)

County	Total poor		Total pop.	% poor		Difference [((b-a)/b)*100]
	Bigman (a)	Spatial p. (b)		Bigman (a)	Spatial p. (b)	
Azogues	34 017	34 051	40 510	84.0	84.1	0.098
Biblian	16 665	16 719	22 286	74.8	75.0	0.323
Cañar	40 194	40 200	44 080	91.2	91.2	0.015
La Troncal	7 628	7 468	9 810	77.8	76.1	-2.141
El Tambo	6 116	6 152	6 974	87.7	88.2	0.582
Deleg	5 431	5 485	7 298	74.4	75.2	0.991
Tulcan	21 574	21 352	26 089	82.7	81.8	-1.038
Bolivar (De Carchi)	12 471	12 442	15 157	82.3	82.1	-0.235
Espejo	9 541	9 469	13 188	72.3	71.8	-0.767
Mira	11 484	11 446	14 040	81.8	81.5	-0.331
Montufar	15 445	15 409	17 481	88.4	88.1	-0.230
San Pedro De Huaca	5 109	5 058	6 485	78.8	78.0	-1.020
Latacunga	63 180	62 597	77 721	81.3	80.5	-0.931
La Mana	9 356	9 271	12 060	77.6	76.9	-0.916
Pangua	13 058	13 149	16 814	77.7	78.2	0.688
Pujili	34 069	34 188	39 038	87.3	87.6	0.348
Salcedo	29 579	29 414	37 740	78.4	77.9	-0.560
Saquisili	9 893	9 857	12 829	77.1	76.8	-0.375
Sigchos	16 943	17 010	18 453	91.8	92.2	0.389
Riobamba	53 423	53 242	69 274	77.1	76.9	-0.340
Alausi	31 583	31 734	33 910	93.1	93.6	0.474
Colta	41 615	42 052	47 568	87.5	88.4	1.038
Chambo	7 636	7 656	9 463	80.7	80.9	0.268
Chunchi	11 660	11 646	13 490	86.4	86.3	-0.119
Guamote	26 917	27 013	28 058	95.9	96.3	0.355
Guano	25 241	25 173	30 522	82.7	82.5	-0.272
Pallatanga	7 864	7 919	9 482	82.9	83.5	0.699
Penipe	5 788	5 730	7 078	81.8	81.0	-1.005
Cumanda	5 711	5 785	7 097	80.5	81.5	1.282
Machala	7 958	7 648	13 410	59.3	57.0	-4.058
Arenillas	4 866	4 754	6 728	72.3	70.7	-2.369
Atahualpa	3 728	3 626	6 142	60.7	59.0	-2.803
Balsas	2 651	2 587	4 066	65.2	63.6	-2.488
Chilla	2 055	2 033	2 767	74.3	73.5	-1.083
El Guabo	14 128	13 888	16 955	83.3	81.9	-1.732
Huaquillas	137	138	174	79.0	79.0	0.085
Marcabeli	3 281	3 254	4 889	67.1	66.6	-0.836
Pasaje	12 641	12 468	16 557	76.3	75.3	-1.391
Piñas	7 954	7 765	11 681	68.1	66.5	-2.429
Portovelo	3 541	3 467	4 573	77.4	75.8	-2.131
Santa Rosa	9 665	9 376	14 465	66.8	64.8	-3.085
Zaruma	12 566	12 331	16 455	76.4	74.9	-1.900
Las Lajas	3 943	3 883	4 866	81.0	79.8	-1.529

(Continued)

Table A.2. (Continued)

County	Total poor		Total pop.	% poor		Difference [((b-a)/b)*100]
	Bigman (a)	Spatial p. (b)		Bigman (a)	Spatial p. (b)	
Esmeraldas	16 344	16 308	22 314	73.2	73.1	-0.218
Eloy Alfaro	22 267	22 237	25 389	87.7	87.6	-0.135
Muisne	18 278	18 095	22 537	81.1	80.3	-1.007
Quininde	45 264	44 412	53 254	85.0	83.4	-1.917
San Lorenzo	9 993	9 977	11 322	88.3	88.1	-0.157
Atacames	15 801	15 980	18 014	87.7	88.7	1.122
Rio Verde	16 317	16 211	19 911	81.9	81.4	-0.649
Guayaquil	4 623	4 548	34 618	13.4	13.1	-1.636
Alfredo Baquerizo Moreno	8 224	8 308	10 274	80.0	80.9	1.014
Balao	4 777	4 738	5 966	80.1	79.4	-0.814
Balzar	16 107	16 293	21 082	76.4	77.3	1.138
Colimes	15 342	15 190	19 232	79.8	79.0	-1.001
Daule	31 395	31 159	39 944	78.6	78.0	-0.756
Duran	2 078	2 074	2 837	73.2	73.1	-0.211
El Empalme	23 407	23 385	31 989	73.2	73.1	-0.095
El Triunfo	5 600	5 464	7 240	77.3	75.5	-2.475
Milagro	15 755	15 285	22 760	69.2	67.2	-3.074
Naranjal	23 138	22 843	23 468	98.6	97.3	-1.294
Naranjito	5 769	5 651	8 487	68.0	66.6	-2.095
Palestina	4 657	4 638	5 883	79.2	78.8	-0.404
Pedro Carbo	9 090	9 088	11 648	78.0	78.0	-0.030
Salinas	4 086	4 004	5 302	77.1	75.5	-2.047
Samborondon	12 897	12 920	16 494	78.2	78.3	0.175
Santa Elena	21 143	21 046	24 338	86.9	86.5	-0.459
Santa Lucia	17 820	17 612	20 812	85.6	84.6	-1.176
Urbina Jado	33 197	33 368	36 606	90.7	91.2	0.512
Yaguachi	17 083	17 031	23 078	74.0	73.8	-0.307
Playas	1 302	1 266	1 946	66.9	65.1	-2.837
Simon Bolivar	13 632	13 591	17 240	79.1	78.8	-0.303
Coronel Marcelino Maridueña	2 984	2 906	4 026	74.1	72.2	-2.695
Lomas De Sargentillo	1 016	1 010	1 205	84.3	83.8	-0.607
Nobol (Piedrahita)	7 928	7 793	10 524	75.3	74.1	-1.732
General Antonio Elizalde (Bucay)	1 280	1 281	1 890	67.7	67.8	0.046
Isidro Ayora	4 705	4 646	5 797	81.2	80.1	-1.265
Ibarra	32 791	32 638	38 502	85.2	84.8	-0.466
Antonio Ante	11 013	11 033	13 345	82.5	82.7	0.182
Cotacachi	24 245	24 068	27 199	89.1	88.5	-0.735
Otavalo	30 603	30 547	34 738	88.1	87.9	-0.184
Pimampiro	14 035	14 020	15 359	91.4	91.3	-0.111
San Miguel De Urcuqui	11 263	11 061	13 736	82.0	80.5	-1.825
Loja	41 273	40 825	50 188	82.2	81.3	-1.099
Calvas	16 884	16 895	19 292	87.5	87.6	0.066
Catamayo	7 475	7 441	8 899	84.0	83.6	-0.466

(Continued)

Table A.2. (Continued)

County	Total poor		Total pop.	% poor		Difference [((b-a)/b)*100]
	Bigman (a)	Spatial p. (b)		Bigman (a)	Spatial p. (b)	
Celica	11 264	11 325	14 329	78.6	79.0	0.538
Chaguarpamba	8 111	8 179	9 850	82.3	83.0	0.838
Espindola	15 830	15 815	18 191	87.0	86.9	-0.095
Gonzanama	13 908	13 822	17 276	80.5	80.0	-0.625
Macara	6 044	6 085	7 249	83.4	83.9	0.674
Paltas	19 594	19 567	23 050	85.0	84.9	-0.136
Puyango	13 298	13 309	16 804	79.1	79.2	0.083
Saraguro	22 972	23 002	26 842	85.6	85.7	0.131
Sozoranga	8 338	8 330	9 729	85.7	85.6	-0.085
Zapotillo	7 830	7 931	10 234	76.5	77.5	1.269
Pindal	6 417	6 470	7 457	86.1	86.8	0.820
Quilanga	4 445	4 463	5 553	80.0	80.4	0.409
Olmedo	4 380	4 409	5 146	85.1	85.7	0.658
Babahoyo	36 858	36 153	48 740	75.6	74.2	-1.949
Baba	22 580	22 500	29 406	76.8	76.5	-0.357
Montalvo	7 868	7 806	11 568	68.0	67.5	-0.791
Pueblo Viejo	19 168	19 364	22 662	84.6	85.4	1.013
Quevedo	10 024	9 902	15 869	63.2	62.4	-1.225
Urdaneta	11 066	10 813	16 719	66.2	64.7	-2.332
Ventanas	27 290	27 362	35 277	77.4	77.6	0.264
Vinces	31 278	31 735	36 722	85.2	86.4	1.439
Palenque	15 977	16 215	18 242	87.6	88.9	1.471
Buena Fe	9 574	9 442	13 892	68.9	68.0	-1.398
Valencia	15 265	15 215	19 357	78.9	78.6	-0.332
Mocache	20 916	21 207	26 380	79.3	80.4	1.372
Portoviejo	37 511	36 237	62 368	60.1	58.1	-3.515
Bolivar (De Manabi)	17 646	17 387	25 458	69.3	68.3	-1.491
Chone	51 870	50 623	73 374	70.7	69.0	-2.464
El Carmen	20 765	20 518	28 779	72.2	71.3	-1.202
Flavio Alfaro	16 385	16 244	23 613	69.4	68.8	-0.868
Jipijapa	30 536	30 272	36 062	84.7	83.9	-0.873
Junin	11 177	11 102	18 243	61.3	60.9	-0.668
Manta	3 617	3 609	4 701	76.9	76.8	-0.214
Montecristi	5 609	5 598	7 211	77.8	77.6	-0.203
Pajan	30 839	30 808	36 957	83.4	83.4	-0.099
Pichincha	21 760	21 474	28 827	75.5	74.5	-1.334
Rocafuerte	13 013	12 891	17 177	75.8	75.0	-0.952
Santa Ana	30 758	30 230	42 796	71.9	70.6	-1.745
Sucre	18 546	18 392	27 685	67.0	66.4	-0.837
Tosagua	17 886	17 648	24 366	73.4	72.4	-1.350
24 De Mayo	23 902	23 969	34 026	70.2	70.4	0.279
Pedernales	19 651	19 364	25 228	77.9	76.8	-1.484
Olmedo	7 402	7 365	10 166	72.8	72.4	-0.505

(Continued)

Table A.2. (Continued)

County	Total poor		Total pop.	% poor		Difference [((b-a)/b)*100]
	Bigman (a)	Spatial p. (b)		Bigman (a)	Spatial p. (b)	
Puerto Lopez	5 052	5 037	5 587	90.4	90.2	-0.293
Jama	7 404	7 376	10 039	73.8	73.5	-0.380
Jaramijo	239	239	298	80.2	80.1	-0.088
Morona	10 197	10 059	12 316	82.8	81.7	-1.373
Gualaquiza	8 254	8 039	12 518	65.9	64.2	-2.679
Limon	5 995	5 949	8 813	68.0	67.5	-0.770
Palanda	3 848	3 764	5 525	69.7	68.1	-2.232
Santiago	6 006	5 906	8 319	72.2	71.0	-1.685
Sucua	7 231	7 033	10 996	65.8	64.0	-2.817
Huamboya	2 976	2 949	3 922	75.9	75.2	-0.916
San Juan Bosco	2 336	2 316	3 249	71.9	71.3	-0.859
Taisha	5 664	5 673	6 666	85.0	85.1	0.146
Logroño	3 150	3 114	3 646	86.4	85.4	-1.143
Tena	20 345	20 140	25 464	79.9	79.1	-1.019
Archidona	11 194	11 027	13 650	82.0	80.8	-1.509
El Chaco	3 077	2 992	4 445	69.2	67.3	-2.863
Quijos	2 329	2 251	3 847	60.5	58.5	-3.464
Carlos Julio Arosemena Tola	1 437	1 428	2 037	70.5	70.1	-0.598
Pastaza	10 469	10 488	15 342	68.2	68.4	0.186
Mera	3 427	3 310	5 947	57.6	55.7	-3.557
Santa Clara	1 631	1 620	2 233	73.0	72.5	-0.712
Arajuno	3 180	3 183	3 594	88.5	88.6	0.085
Quito	6 315	3 760	169 507	3.7	2.2	-67.958
Cayambe	20 858	20 342	29 848	69.9	68.2	-2.537
Mejia	19 876	18 841	37 308	53.3	50.5	-5.499
Pedro Moncayo	7 377	7 047	15 718	46.9	44.8	-4.686
Rumiñahui	4 312	4 208	8 620	50.0	48.8	-2.468
Sto Domingo De Los Colorados	35 840	33 865	68 261	52.5	49.6	-5.831
San Miguel De Los Bancos	10 449	10 442	16 928	61.7	61.7	-0.062
Pedro Vicente Maldonado	4 669	4 511	7 681	60.8	58.7	-3.494
Puerto Quito	10 086	9 351	13 775	73.2	67.9	-7.860
Ambato	60 766	59 913	103 624	58.6	57.8	-1.423
Baños	4 059	3 942	5 915	68.6	66.6	-2.961
Cevallos	3 452	3 317	5 964	57.9	55.6	-4.074
Mocha	4 329	4 212	6 368	68.0	66.2	-2.776
Patate	5 932	5 756	10 292	57.6	55.9	-3.072
Quero	11 086	10 812	15 997	69.3	67.6	-2.531
Pelileo	20 153	19 495	31 665	63.6	61.6	-3.376
Pillaro	18 096	17 724	28 223	64.1	62.8	-2.099
Tisaleo	6 599	6 452	9 165	72.0	70.4	-2.270
Zamora	10 147	9 967	13 695	74.1	72.8	-1.805
Chinchipe	5 243	5 218	6 989	75.0	74.7	-0.483
Nangaritza	3 137	3 111	4 316	72.7	72.1	-0.827

(Continued)

Table A.2. (Continued)

County	Total poor		Total pop.	% poor		Difference [((b-a)/b)*100]
	Bigman (a)	Spatial p. (b)		Bigman (a)	Spatial p. (b)	
Yacuambi	2 751	2 766	3 448	79.8	80.2	0.542
Yantzaza	7 775	7 669	11 856	65.6	64.7	-1.385
El Pangui	4 295	4 274	6 054	70.9	70.6	-0.509
Centinela Del Condor	4 504	4 450	6 331	71.1	70.3	-1.199
Palanda	4 513	4 483	5 430	83.1	82.6	-0.669
Lago Agrio	20 257	20 015	25 588	79.2	78.2	-1.208
Gonzalo Pizarro	3 290	3 206	4 472	73.6	71.7	-2.615
Putumayo	4 145	4 122	4 794	86.5	86.0	-0.548
Shushufindi	14 736	14 462	18 977	77.7	76.2	-1.899
Sucumbios	2 090	2 049	2 441	85.6	84.0	-2.004
Cascales	3 719	3 644	5 014	74.2	72.7	-2.067
Cuyabeno	2 121	2 098	2 501	84.8	83.9	-1.061
Orellana	9 075	9 055	11 369	79.8	79.6	-0.223
Aguarico	2 733	2 738	3 150	86.8	86.9	0.196
La Joya De Los Sachas	12 571	12 458	16 193	77.6	76.9	-0.911
Loreto	6 915	6 897	7 811	88.5	88.3	-0.251
Total	**2 875 141**	**2 847 776**	**3 975 625**	**72.3**	**71.6**	**-0.977**

Figure A.2. **Percentage of poor people per county, rural sample**

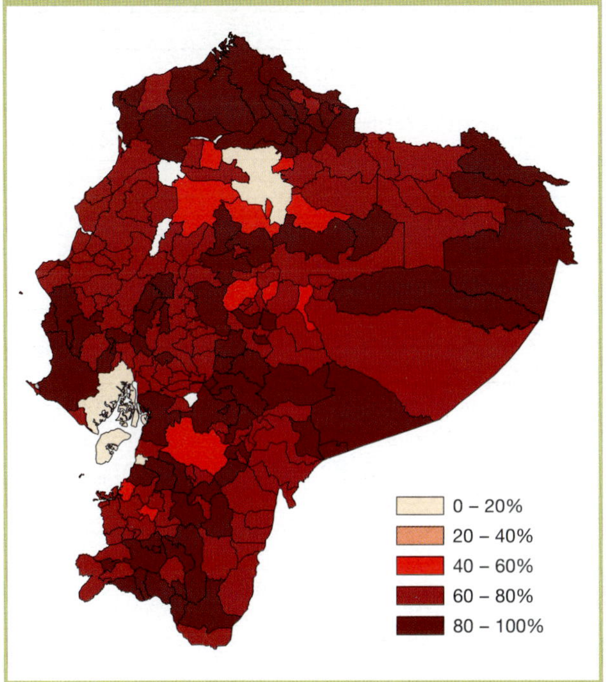

Figure A.3. **Comparison between probit method and spatial probit at county level, rural sample**

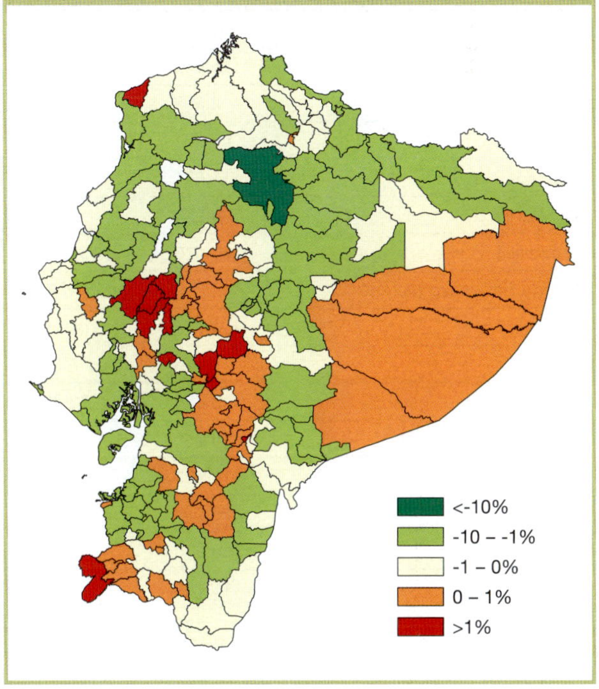

Table A.3. **Comparison between probit and spatial probit in the headcount index and in the percentage of poor people at county level, urban sample**

County	Total poor		Total pop.	% poor		Difference [((b-a)/b)*100]
	Bigman (a)	Spatial p. (b)		Bigman (a)	Spatial p. (b)	
Cuenca	23 672	26 660	194 981	12.1	13.7	11.208
Gualaceo	4 111	4 393	8 572	48.0	51.2	6.423
Guaranda	4 566	5 029	15 730	29.0	32.0	9.198
Azogues	7 830	8 462	21 060	37.2	40.2	7.468
Cañar	6 412	6 810	11 179	57.4	60.9	5.851
La Troncal	14 510	14 944	22 730	63.8	65.7	2.905
Tulcan	16 040	17 245	37 069	43.3	46.5	6.993
Montufar	5 849	6 276	11 973	48.9	52.4	6.804
Latacunga	11 880	12 533	39 882	29.8	31.4	5.204
La Mana	5 077	5 293	8 900	57.0	59.5	4.074
Pujili	1 274	1 366	5 305	24.0	25.7	6.704
Salcedo	2 220	2 394	7 582	29.3	31.6	7.249
Riobamba	19 657	21 753	94 505	20.8	23.0	9.635
Alausi	2 526	2 739	5 399	46.8	50.7	7.793
Guano	3 251	3 480	6 584	49.4	52.9	6.584
Machala	41 533	42 799	144 197	28.8	29.7	2.959
Arenillas	4 836	5 055	11 586	41.7	43.6	4.347
El Guabo	5 743	5 929	11 103	51.7	53.4	3.151
Huaquillas	17 349	17 848	27 632	62.8	64.6	2.791
Pasaje	16 024	17 069	34 849	46.0	49.0	6.124
Piñas	3 732	3 958	10 162	36.7	38.9	5.718
Portovelo	2 271	2 371	5 684	40.0	41.7	4.242
Santa Rosa	16 261	17 149	36 387	44.7	47.1	5.181
Zaruma	2 298	2 472	7 397	31.1	33.4	7.020
Esmeraldas	41 221	45 619	112 831	36.5	40.4	9.640
Quininde	11 871	12 096	18 647	63.7	64.9	1.865
San Lorenzo	8 050	8 375	11 230	71.7	74.6	3.881
Guayaquil	333 578	493 451	1 535 393	21.7	32.1	32.399
Alfredo Baquerizo Moreno	4 009	4 303	5 979	67.1	72.0	6.832
Balao	4 792	4 992	6 548	73.2	76.2	4.002
Balzar	14 416	15 517	22 883	63.0	67.8	7.096
Daule	13 076	13 829	25 590	51.1	54.0	5.446
Duran	29 086	32 004	82 359	35.3	38.9	9.119
El Empalme	14 791	15 807	26 362	56.1	60.0	6.433
El Triunfo	10 275	10 601	17 311	59.4	61.2	3.072
Milagro	39 938	42 203	93 637	42.7	45.1	5.365
Naranjal	15 084	14 745	15 998	94.3	92.2	-2.302
Naranjito	7 688	8 155	17 121	44.9	47.6	5.724
Palestina	3 012	3 191	5 308	56.7	60.1	5.628
Pedro Carbo	13 037	13 715	19 980	65.3	68.6	4.940
Salinas	12 425	13 037	27 132	45.8	48.1	4.694
Samborondon	6 051	6 709	17 471	34.6	38.4	9.808

(Continued)

Table A.3. (Continued)

County	Total poor		Total pop.	% poor		Difference [((b-a)/b)*100]
	Bigman (a)	Spatial p. (b)		Bigman (a)	Spatial p. (b)	
Santa Elena	38 401	40 159	59 672	64.4	67.3	4.377
Santa Lucia	4 186	4 310	6 530	64.1	66.0	2.878
Urbina Jado	5 265	5 536	7 206	73.1	76.8	4.890
Yaguachi	9 486	10 061	16 245	58.4	61.9	5.714
Playas	11 167	11 636	19 544	57.1	59.5	4.032
Lomas De Sargentillo	6 336	6 569	9 205	68.8	71.4	3.546
La Libertad	31 389	32 763	53 108	59.1	61.7	4.193
General Antonio Elizalde (Bucay)	2 172	2 373	5 078	42.8	46.7	8.499
Ibarra	35 140	37 695	80 991	43.4	46.5	6.779
Antonio Ante	7 146	7 769	14 030	50.9	55.4	8.028
Cotacachi	2 669	2 787	6 051	44.1	46.1	4.217
Otavalo	9 542	9 948	21 548	44.3	46.2	4.076
Loja	25 907	27 755	94 305	27.5	29.4	6.660
Calvas	4 114	4 478	10 106	40.7	44.3	8.134
Catamayo	8 251	8 692	13 458	61.3	64.6	5.076
Macara	5 654	6 186	11 032	51.3	56.1	8.600
Paltas	2 397	2 556	5 555	43.2	46.0	6.208
Babahoyo	20 220	21 291	56 731	35.6	37.5	5.031
Montalvo	2 276	2 430	7 455	30.5	32.6	6.356
Quevedo	32 225	34 638	88 306	36.5	39.2	6.968
Urdaneta	3 158	3 274	6 740	46.8	48.6	3.561
Ventanas	12 665	13 587	23 217	54.6	58.5	6.786
Vinces	8 839	9 797	17 512	50.5	55.9	9.779
Buena Fe	9 774	10 253	19 478	50.2	52.6	4.678
Valencia	4 207	4 468	7 913	53.2	56.5	5.854
Mocache	3 031	3 301	5 373	56.4	61.4	8.170
Portoviejo	29 543	31 300	139 493	21.2	22.4	5.613
Bolivar (De Manabi)	4 537	4 872	12 122	37.4	40.2	6.880
Chone	14 188	15 088	41 437	34.2	36.4	5.961
El Carmen	12 908	13 570	25 291	51.0	53.7	4.882
Jipijapa	18 434	19 180	33 115	55.7	57.9	3.894
Manta	43 404	46 247	128 115	33.9	36.1	6.148
Montecristi	13 399	14 203	22 425	59.7	63.3	5.665
Pajan	3 320	3 503	5 489	60.5	63.8	5.219
Rocafuerte	3 476	3 725	8 844	39.3	42.1	6.679
Santa Ana	2 218	2 362	6 368	34.8	37.1	6.083
Sucre	16 865	18 141	37 637	44.8	48.2	7.033
Tosagua	3 050	3 243	7 412	41.1	43.8	5.981
Pedernales	3 160	3 280	5 449	58.0	60.2	3.656
Puerto Lopez	5 548	5 745	8 043	69.0	71.4	3.443
Jaramijo	5 182	5 412	7 726	67.1	70.1	4.261
Morona	2 996	3 160	8 246	36.3	38.3	5.169
Tena	2 052	2 225	7 873	26.1	28.3	7.774

(Continued)

Table A.3. (Continued)

County	Total poor		Total pop.	% poor		Difference [((b-a)/b)*100]
	Bigman (a)	Spatial p. (b)		Bigman (a)	Spatial p. (b)	
Pastaza	2 520	2 848	14 438	17.5	19.7	11.505
Quito	98 089	144 878	1 201 954	8.2	12.1	32.296
Cayambe	4 132	4 204	17 090	24.2	24.6	1.707
Mejia	2 312	2 397	9 379	24.6	25.6	3.574
Rumiñahui	5 704	6 507	37 595	15.2	17.3	12.347
Sto Domingo De Los Colorados	34 305	35 769	122 564	28.0	29.2	4.092
Ambato	18 040	20 220	124 166	14.5	16.3	10.781
Baños	2 919	3 085	9 501	30.7	32.5	5.362
Pelileo	1 221	1 273	5 954	20.5	21.4	4.091
Pillaro	1 033	1 111	5 146	20.1	21.6	7.006
Zamora	2 405	2 579	8 048	29.9	32.1	6.776
Orellana	3 551	3 781	7 805	45.5	48.4	6.098
Total	**1 497 815**	**1 771 445**	**5 597 559**	**26.7**	**31.6**	**15.506**

Figure A.4. **Percentage of poor people per county, urban sample**

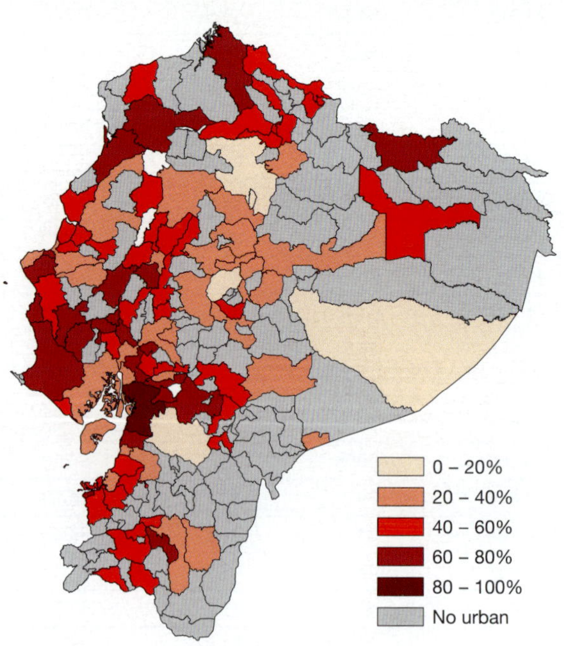

Figure A.5. **Comparison between probit method and spatial probit at county level, urban sample**

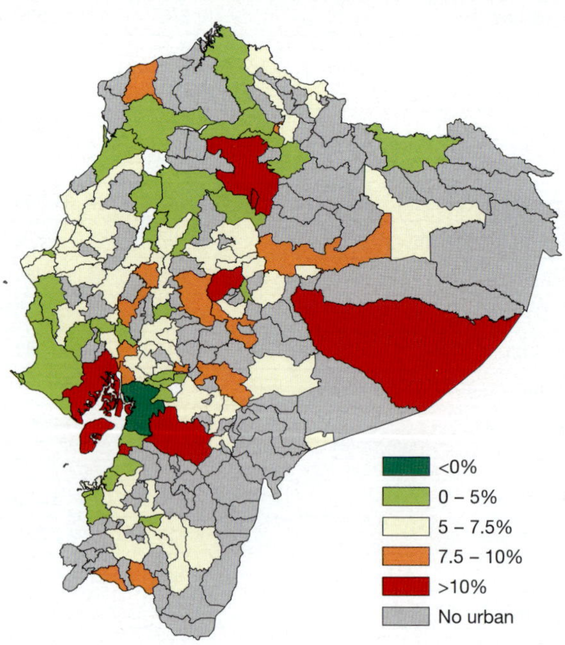

Figure A.6. **Comparison between probit method and spatial probit at province level, total sample**

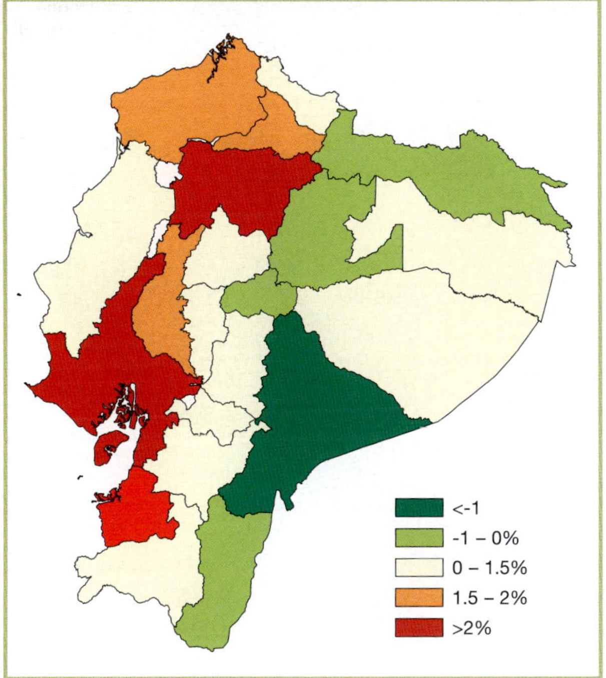

Figure A.7. **Percentage of poor people per province, rural sample**

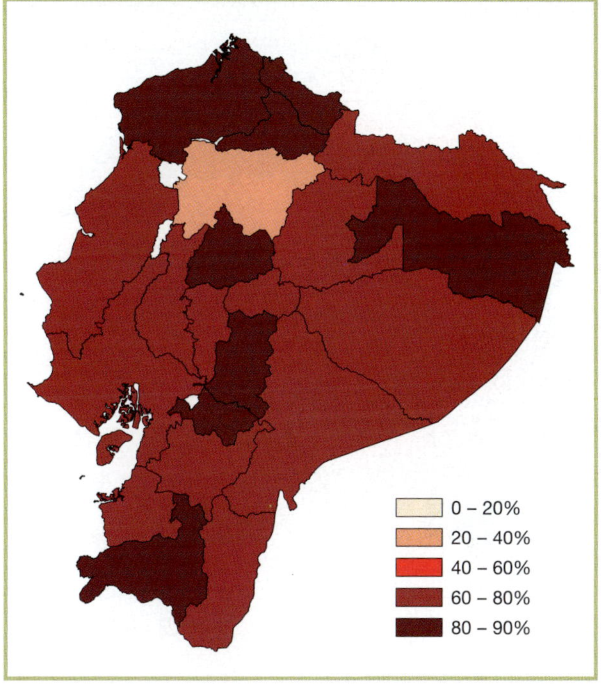

Figure A.8. **Comparison between probit method and spatial probit at province level, rural sample**

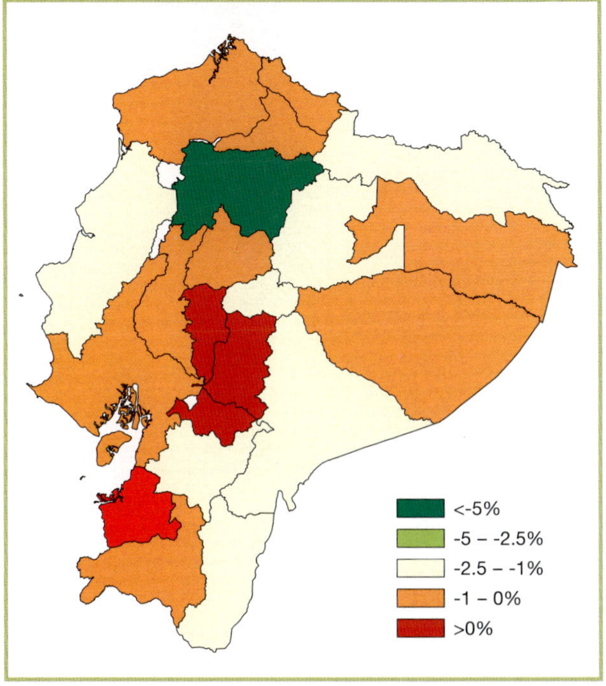

Figure A.9. **Percentage of poor people per province, urban sample**

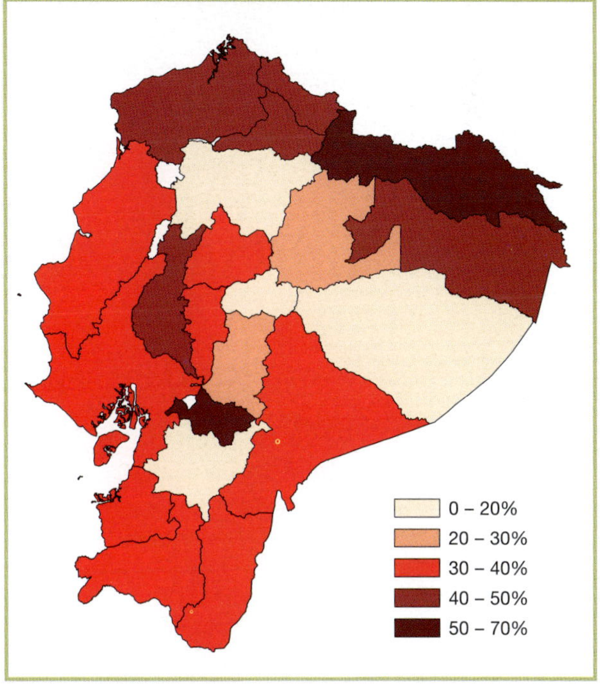

Figure A.10. **Comparison between probit method and spatial probit at province level, urban sample**

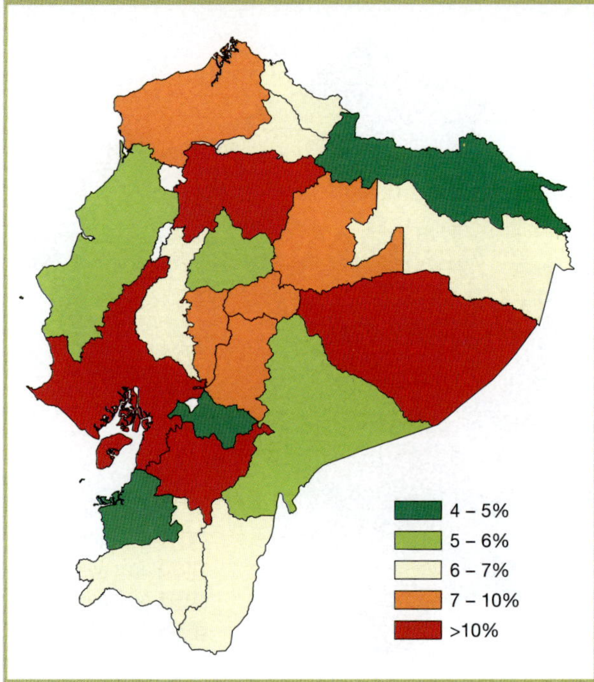

Figure A.11. **Percentage of poor people per region**

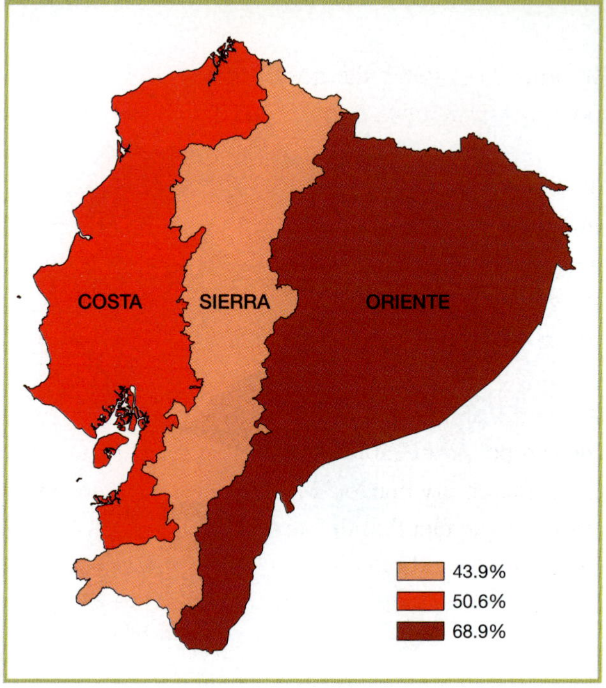

Annex B. **Changing the definition of poverty line**

In order to evaluate the implications of using different measures of poverty, another indicator of poverty was applied to the same data, using the same statistical analysis applied before.

The new poverty line is different from the one adopted in Chapter 5 as it considers the possible influence on the household's consumption of living in different areas of Ecuador. In particular, the average of the households' consumptions at the area level (urban or rural) for each region of Ecuador (Sierra, Oriente and Costa) was calculated. Finally, the new poverty line was considered to be two-thirds of the average of the six average consumptions obtained.

A comparison of the results obtained from adopting this last measure of poverty (Table B.1) with those reported in Table 7 shows that the percentage of poor people decreases from 48 percent with the vulnerability line to 34 percent with the new poverty line. One possible reason for such different results is the fact that the new poverty line includes the area dimension. This seems to have a real impact on household consumption. Another reason is that the new poverty line is 'higher' than the original one adopted in the study and so fewer households are considered poor.

Table B.1. **Percentage of poor people in the provinces of Ecuador, total sample**

Province	% poor		Difference [((a-b)/a)*100]
	Spatial p. I.2 (a)	Spatial p. I.1(b)	
Azuay	26.6	46.6	-75.188
Bolivar	32.0	70.1	-119.063
Cañar	53.1	75.5	-42.185
Carchi	46.0	69.8	-51.739
Chimboraz	45.0	67.9	-50.889
Cotopaxi	45.5	71.3	-56.703
El Oro	39.5	48.9	-23.797
Esmeralda	35.6	66.4	-86.517
Guayas	43.2	44.3	-2.546
Imbabura	57.6	68.4	-18.750
Loja	45.6	67.0	-46.930
Los Rios	34.0	62.7	-84.412
Manabi	34.3	55.8	-62.682
Morona Sa	31.6	68.8	-117.722
Napo	34.2	69.9	-104.386
Orellana	38.4	75.4	-96.354
Pastaza	30.8	51.6	-67.532
Pichincha	13.6	17.4	-27.941
Sucumbios	41.6	75.2	-80.769
Tungurahu	27.0	43.5	-61.111
Zamora Ch	25.4	67.3	-164.961
Total	**34.7**	**48.2**	**-38.905**

Figure B.1. **Percentage of poor people in the counties of Ecuador, total sample**

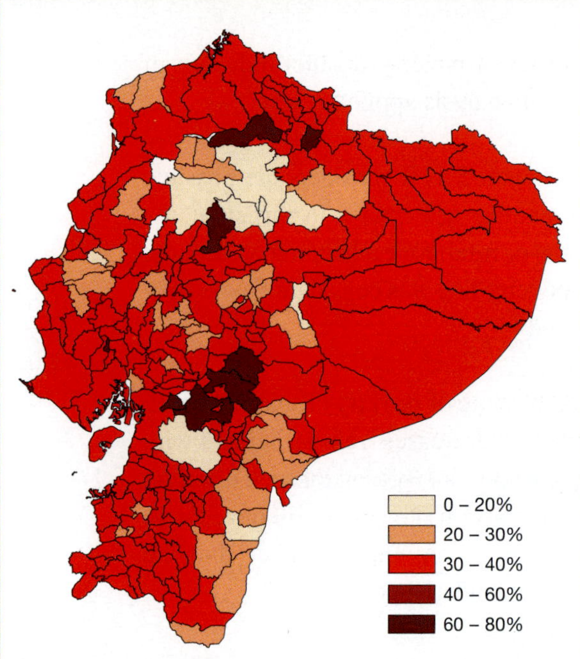

Figure B.2. **Percentage of poor people in the provinces of Ecuador, total sample**

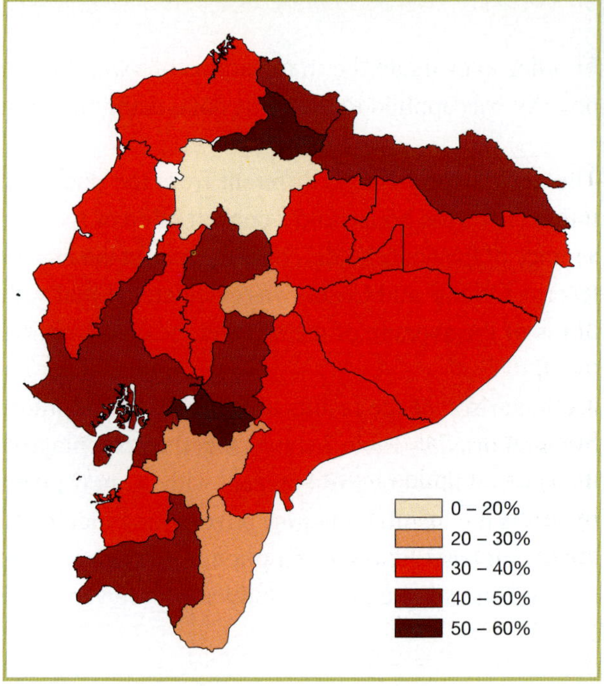

Annex C. **R programmes**

The study used the software tool R for the spatial statistical analysis, and the program codes given in Figure C.1 were used to calculate the spatial correlation and the spatial probit.

The R language was implemented instead of other statistical software (e.g. S-plus) for two main reasons: (i) R is a freeshare program; and (ii) R allows spatial correlation to be calculated on large samples such as the ECV survey covering about 5 600 households. Moreover, R is an open source language (available at: http://cran.r-project.org/bin) created in 1997 by the Department of Statistics of the University of Auckland, New Zealand. R is an integrated suite of software facilities for data manipulation, calculation and graphical display. It has an effective data handling, storage facilities, a suite of operators for calculation on arrays and matrices, and other specific tools for data analysis. Some of the classical and modern statistical techniques are built into the base R environment, and many are supplied in the form of downloadable packages. R can be extended by libraries written by users; and from it, one can use libraries of other languages such as C or Fortran. Users can also create graphic interfaces.

Figure C.1. **R program to calculate Moran and Geary coefficients and spatial probit**

```
##############Spatial Probit##########################################################
rm(list=ls(all=TRUE))
        campc<-read.table("C:/Documents and
Settings//Desktop/ECVfamiglievaramb.txt",col.names=c("ID","xfam","yfam","Analfa","Istruzsup","Region",
"Piso","acqua","servig","pareti","elettricit","basura","persc","TOTPER","TASAM","BEBES","Erosione",
"Clima","Inondazioni","Vulcani","Areacod","REXPLRGE","Plinefamiglia","yprob","road1","road2","road3",
"TOTAREA","Produttivit","PROT","IRR","Foreste","Arable"))
coord<-cbind(campc$xfam,campc$yfam)
coord<-coord/1000
cons<-log((campc$REXPLRGE))
n<-5630
media<-mean(cons)
dev<-var(cons)*n
diffm<-0
diffg<-0
k<-matrix(0,5630,1)
s<-matrix(0,5630,1)
s1<-0
ystar<-matrix(0,5630,1)
vic<-matrix(0,1,5630)
for (i in 1:5630) {for (j in 1:5630)
        {dist<-sqrt((coord[i,1]-coord[j,1])^2+(coord[i,2]-coord[j,2])^2);
        if (i!=j) {if (dist<=40) {vic[j]<-1/dist
            diffm<-diffm+(((cons[i]-media)*(cons[j]-media))*vic[j]/dev)
            diffg<-diffg+(((cons[i]-cons[j])^2)/dev)
             k[j]<-k[j]+(1/dist)
             s[i]<-s[i]+1/dist
             s1<-s1+((1/dist)+(1/dist))^2}
            else vic[j]<-0}
          else vic[j]<-0}
        ystar[i,1]<-vic%*%campc$yprob;
        vic<-0}
a<-sum(k)
Moran<-(n/(a))*diffm
Geary<-(n-1)/(2*a)*diffg
s0<-sum(k)
s1<-(1/2)*s1
```

(Continued)

Figure C.1. (Continued)

```
s2<-0
m4<-sum((cons-media)^4)/n
k1<-m4/((dev/n)^2)
for (i in 1:5630) {s2<-s2+(s[i]+k[i])^2}
varM<-((n*(((n)^2-3*n+3)*s1-n*s2+3*(s0)^2)-k1*(n*(n-1)*s1-2*n*s2+6*s0^2))/((n-1)*(n-2)*(n-3)*s0^2))-(1/(n-1)^2)
z<-(Moran-(-1/(n-1)))/sqrt(varM)
probabi<-dnorm(z)
ris<-c("Moran=",Moran)
if (probabi<0.05) ris1<-c("Rifiuto H0: Not AUTOCORRELATION") else ris1<-c("Accept H0: NON
 AUTOCORRELAZIONE")
ris2<-c("Geary=",Geary)
risultati<-cbind(ris,ris1,ris2)
write.table(risultati,file="C:/Documents and Settings/Desktop/LMorant.txt")
write.table(ystar,file="C:/Documents and Settings/Desktop/tystar.txt")
ystar<-matrix(scan("c:/Documents and Settings/Desktop/tystar.txt"),ncol=1,byrow=T)
campc<-data.frame(campc,ystar)
campc$Erosione<-factor(campc$Erosione)
campc$Clima<-factor(campc$Clima)
campc$Arable<-factor(campc$Arable)
Spprobittot<-(glm(yprob~1+Analfa+Istruzsup+Piso+
    acqua+servig+pareti+elettricit+basura+persc+TOTPER+TASAM+BEBES+Erosione+
Clima+Inondazioni+Vulcani+ystar+Areacod+road1+road2+road3+
TOTAREA+Produttivit+PROT+IRR+Foreste+Arable,data=campc,family=binomial(link="probit")))
deviance(Spprobittot)
summary(Spprobittot)
anova(Spprobittot,test="Chisq")
win.graph()
par(mfrow=c(2,2))
plot(resid(Spprobittot,type="response"),xlab="Y",ylab="ei")
plot(resid(Spprobittot,type="pearson"),xlab="Y",ylab="epi")
plot(resid(Spprobittot,type="deviance"),xlab="Y",ylab="edi")
betapros<-c(Spprobittot$coef[1:21],Spprobittot$coef[23:33])
x<-matrix(scan("c:/Documents and Settings/salvati/Desktop/xrural.txt"),ncol=32,byrow=T)
yprob<-x%*%betapros
povs<-pnorm(yprob)
plot(povs)
write.table(povs,file="C:/Documents and Settings/salvati/Desktop/spatprobruramb.txt")

##########################Probit (Bigman)##############################################
rm(list=ls(all=TRUE))
    campc<-read.table("C:/Documents and
Settings/Desktop/ECVfamiglievaramb.txt",col.names=c("ID","xfam","yfam","Analfa","Istruzsup","Region",

"Piso","acqua","servig","pareti","elettricit","basura","persc","TOTPER","TASAM","BEBES","Erosione",
"Clima","Inondazioni","Vulcani","Areacod","REXPLRGE","Plinefamiglia","yprob","road1","road2","road3",
"TOTAREA","Produttivit","PROT","IRR","Foreste","Arable"))

campc$Erosione<-factor(campc$Erosione)
campc$Clima<-factor(campc$Clima)
campc$Arable<-factor(campc$Arable)

probittot<-(glm(yprob~1+Analfa+Istruzsup+Piso+
    acqua+servig+pareti+elettricit+basura+persc+TOTPER+TASAM+BEBES+Erosione+
Clima+Inondazioni+Vulcani+Areacod+road1+road2+road3+
TOTAREA+Produttivit+PROT+IRR+Foreste+Arable,data=campc,family=binomial(link="probit")))

deviance(probittot)
summary(probittot)
anova(probittot,test="Chisq")
win.graph()
par(mfrow=c(2,2))
plot(resid(probittot,type="response"),xlab="Y",ylab="ei")
plot(resid(probittot,type="pearson"),xlab="Y",ylab="epi")
plot(resid(probittot,type="deviance"),xlab="Y",ylab="edi")
beta<-probittot$coef
x<-matrix(scan("c:/Documents and Settings/Desktop/xurban.txt"),ncol=32,byrow=T)
yprob<-x%*%beta
povs<-pnorm(yprob)
plot(povs)
write.table(povs,file="C:/Documents and Settings/Desktop/probruramb.txt")
```

Annex D. **Maps and tables of administrative borders and environmental variables**

Figure D.1. **Percentage of poor people in the provinces of Ecuador, total sample**

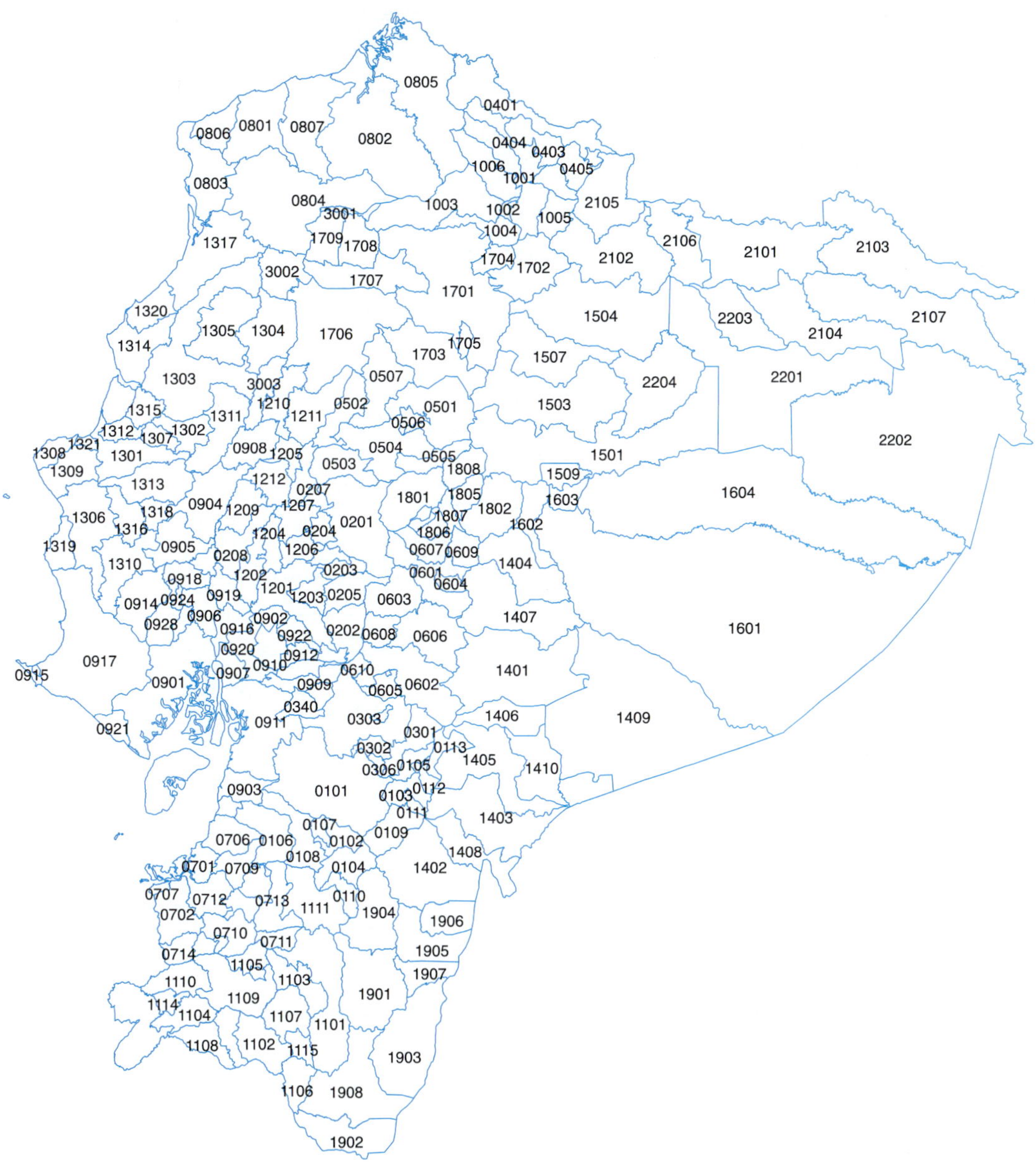

Table D.1. **County codes, Equador**

County code	County description	Province code	Province description	Region
0101	Cuenca	01	Azuay	Sierra
0102	Giron	01	Azuay	Sierra
0103	Gualaceo	01	Azuay	Sierra
0104	Nabon	01	Azuay	Sierra
0105	Paute	01	Azuay	Sierra
0106	Pucara	01	Azuay	Sierra
0107	San Fernando	01	Azuay	Sierra
0108	Santa Isabel	01	Azuay	Sierra
0109	Sigsig	01	Azuay	Sierra
0110	Oña	01	Azuay	Sierra
0111	Chordeleg	01	Azuay	Sierra
0112	El Pan	01	Azuay	Sierra
0113	Sevilla De Oro	01	Azuay	Sierra
0114	Guachapala	01	Azuay	Sierra
0201	Guaranda	02	Bolivar	Sierra
0202	Chillanes	02	Bolivar	Sierra
0203	Chimbo	02	Bolivar	Sierra
0204	Echeandia	02	Bolivar	Sierra
0205	San Miguel	02	Bolivar	Sierra
0206	Caluma	02	Bolivar	Sierra
0207	Las Naves	02	Bolivar	Sierra
0301	Azogues	03	Cañar	Sierra
0302	Biblian	03	Cañar	Sierra
0303	Cañar	03	Cañar	Sierra
0304	La Troncal	03	Cañar	Sierra
0305	El Tambo	03	Cañar	Sierra
0306	Deleg	03	Cañar	Sierra
0401	Tulcan	04	Carchi	Sierra
0402	Bolivar	04	Carchi	Sierra
0403	Espejo	04	Carchi	Sierra
0404	Mira	04	Carchi	Sierra
0405	Montufar	04	Carchi	Sierra
0406	San Pedro De Huaca	04	Carchi	Sierra
0501	Latacunga	05	Cotopaxi	Sierra
0502	La Mana	05	Cotopaxi	Sierra
0503	Pangua	05	Cotopaxi	Sierra
0504	Pujili	05	Cotopaxi	Sierra
0505	Salcedo	05	Cotopaxi	Sierra
0506	Saquisili	05	Cotopaxi	Sierra
0507	Sigchos	05	Cotopaxi	Sierra
0601	Riobamba	06	Chimborazo	Sierra
0602	Alausi	06	Chimborazo	Sierra
0603	Colta	06	Chimborazo	Sierra
0604	Chambo	06	Chimborazo	Sierra

(Continued)

Table D.1. (Continued)

County code	County description	Province code	Province description	Region
0605	Chunchi	06	Chimborazo	Sierra
0606	Guamote	06	Chimborazo	Sierra
0607	Guano	06	Chimborazo	Sierra
0608	Pallatanga	06	Chimborazo	Sierra
0609	Penipe	06	Chimborazo	Sierra
0610	Cumanda	06	Chimborazo	Sierra
0701	Machala	07	El Oro	Costa
0702	Arenillas	07	El Oro	Costa
0703	Atahualpa	07	El Oro	Costa
0704	Balsas	07	El Oro	Costa
0705	Chilla	07	El Oro	Costa
0706	El Guabo	07	El Oro	Costa
0707	Huaquillas	07	El Oro	Costa
0708	Marcabeli	07	El Oro	Costa
0709	Pasaje	07	El Oro	Costa
0710	Piñas	07	El Oro	Costa
0711	Portovelo	07	El Oro	Costa
0712	Santa Rosa	07	El Oro	Costa
0713	Zaruma	07	El Oro	Costa
0714	Las Lajas	07	El Oro	Costa
0801	Esmeraldas	08	Esmeraldas	Costa
0802	Eloy Alfaro	08	Esmeraldas	Costa
0803	Muisne	08	Esmeraldas	Costa
0804	Quininde	08	Esmeraldas	Costa
0805	San Lorenzo	08	Esmeraldas	Costa
0806	Atacames	08	Esmeraldas	Costa
0807	Rio Verde	08	Esmeraldas	Costa
0901	Guayaquil	09	Guayas	Costa
0902	Alfredo Baquerizo Moreno (Juja)	09	Guayas	Costa
0903	Balao	09	Guayas	Costa
0904	Balzar	09	Guayas	Costa
0905	Colimes	09	Guayas	Costa
0906	Daule	09	Guayas	Costa
0907	Duran	09	Guayas	Costa
0908	El Empalme	09	Guayas	Costa
0909	El Triunfo	09	Guayas	Costa
0910	Milagro	09	Guayas	Costa
0911	Naranjal	09	Guayas	Costa
0912	Naranjito	09	Guayas	Costa
0913	Palestina	09	Guayas	Costa
0914	Pedro Carbo	09	Guayas	Costa
0915	Salinas	09	Guayas	Costa
0916	Samborondon	09	Guayas	Costa
0917	Santa Elena	09	Guayas	Costa

(Continued)

Table D.1. (Continued)

County code	County description	Province code	Province description	Region
0918	Santa Lucia	09	Guayas	Costa
0919	Urbina Jado	09	Guayas	Costa
0920	Yaguachi	09	Guayas	Costa
0921	Playas	09	Guayas	Costa
0922	Simon Bolivar	09	Guayas	Costa
0923	Coronel Marcelino Maridueña	09	Guayas	Costa
0924	Lomas De Sargentillo	09	Guayas	Costa
0925	Nobol	09	Guayas	Costa
0926	La Libertad	09	Guayas	Costa
0927	General Antonio Elizalde	09	Guayas	Costa
0928	Isidro Ayora	09	Guayas	Costa
1001	Ibarra	10	Imbabura	Sierra
1002	Antonio Ante	10	Imbabura	Sierra
1003	Cotacachi	10	Imbabura	Sierra
1004	Otavalo	10	Imbabura	Sierra
1005	Pimampiro	10	Imbabura	Sierra
1006	San Miguel De Urcuqui	10	Imbabura	Sierra
1101	Loja	11	Loja	Sierra
1102	Calvas	11	Loja	Sierra
1103	Catamayo	11	Loja	Sierra
1104	Celica	11	Loja	Sierra
1105	Chaguarpamba	11	Loja	Sierra
1106	Espindola	11	Loja	Sierra
1107	Gonzanama	11	Loja	Sierra
1108	Macara	11	Loja	Sierra
1109	Paltas	11	Loja	Sierra
1110	Puyango	11	Loja	Sierra
1111	Saraguro	11	Loja	Sierra
1112	Sozoranga	11	Loja	Sierra
1113	Zapotillo	11	Loja	Sierra
1114	Pindal	11	Loja	Sierra
1115	Quilanga	11	Loja	Sierra
1116	Olmedo	11	Loja	Sierra
1201	Babahoyo	12	Los Rios	Costa
1202	Baba	12	Los Rios	Costa
1203	Montalvo	12	Los Rios	Costa
1204	Puebloviejo	12	Los Rios	Costa
1205	Quevedo	12	Los Rios	Costa
1206	Urdaneta	12	Los Rios	Costa
1207	Ventanas	12	Los Rios	Costa
1208	Vinces	12	Los Rios	Costa
1209	Palenque	12	Los Rios	Costa
1210	Buena Fe	12	Los Rios	Costa
1211	Valencia	12	Los Rios	Costa

(Continued)

Table D.1. (Continued)

County code	County description	Province code	Province description	Region
1212	Mocache	12	Los Rios	Costa
1301	Portoviejo	13	Manabi	Costa
1302	Bolivar	13	Manabi	Costa
1303	Chone	13	Manabi	Costa
1304	El Carmen	13	Manabi	Costa
1305	Flavio Alfaro	13	Manabi	Costa
1306	Jipijapa	13	Manabi	Costa
1307	Junin	13	Manabi	Costa
1308	Manta	13	Manabi	Costa
1309	Montecristi	13	Manabi	Costa
1310	Pajan	13	Manabi	Costa
1311	Pichincha	13	Manabi	Costa
1312	Rocafuerte	13	Manabi	Costa
1313	Santa Ana	13	Manabi	Costa
1314	Sucre	13	Manabi	Costa
1315	Tosagua	13	Manabi	Costa
1316	24 De Mayo	13	Manabi	Costa
1317	Pedernales	13	Manabi	Costa
1318	Olmedo	13	Manabi	Costa
1319	Puerto Lopez	13	Manabi	Costa
1320	Jama	13	Manabi	Costa
1321	Jaramijo	13	Manabi	Costa
1401	Morona	14	Morona Santiago	Oriente
1402	Gualaquiza	14	Morona Santiago	Oriente
1403	Limon Indanza	14	Morona Santiago	Oriente
1404	Palanda	14	Morona Santiago	Oriente
1405	Santiago	14	Morona Santiago	Oriente
1406	Sucua	14	Morona Santiago	Oriente
1407	Huamboya	14	Morona Santiago	Oriente
1408	San Juan Bosco	14	Morona Santiago	Oriente
1409	Taisha	14	Morona Santiago	Oriente
1410	Logroño	14	Morona Santiago	Oriente
1501	Tena	15	Napo	Oriente
1503	Archidona	15	Napo	Oriente
1504	El Chaco	15	Napo	Oriente
1507	Quijos	15	Napo	Oriente
1509	Carlos Julio Arosemena Tola	15	Napo	Oriente
1601	Pastaza	16	Pastaza	Oriente
1602	Mera	16	Pastaza	Oriente
1603	Santa Clara	16	Pastaza	Oriente
1604	Arajuno	16	Pastaza	Oriente
1701	Quito	17	Pichincha	Sierra
1702	Cayambe	17	Pichincha	Sierra
1703	Mejia	17	Pichincha	Sierra

(Continued)

Table D.1. (Continued)

County code	County description	Province code	Province description	Region
1704	Pedro Moncayo	17	Pichincha	Sierra
1705	Rumiñahui	17	Pichincha	Sierra
1706	Santo Domingo	17	Pichincha	Sierra
1707	San Miguel De Los Bancos	17	Pichincha	Sierra
1708	Pedro Vicente Maldonado	17	Pichincha	Sierra
1709	Puerto Quito	17	Pichincha	Sierra
1801	Ambato	18	Tungurahua	Sierra
1802	Baños	18	Tungurahua	Sierra
1803	Cevallos	18	Tungurahua	Sierra
1804	Mocha	18	Tungurahua	Sierra
1805	Patate	18	Tungurahua	Sierra
1806	Quero	18	Tungurahua	Sierra
1807	Pelileo	18	Tungurahua	Sierra
1808	Pillaro	18	Tungurahua	Sierra
1809	Tisaleo	18	Tungurahua	Sierra
1901	Zamora	19	Zamora Chinchipe	Oriente
1902	Chinchipe	19	Zamora Chinchipe	Oriente
1903	Nangaritza	19	Zamora Chinchipe	Oriente
1904	Yacuambi	19	Zamora Chinchipe	Oriente
1905	Yantzaza	19	Zamora Chinchipe	Oriente
1906	El Pangui	19	Zamora Chinchipe	Oriente
1907	Centinela Del Condor	19	Zamora Chinchipe	Oriente
1908	Palanda	19	Zamora Chinchipe	Oriente
2101	Lago Agrio	21	Sucumbios	Oriente
2102	Gonzalo Pizarro	21	Sucumbios	Oriente
2103	Putumayo	21	Sucumbios	Oriente
2104	Shushufindi	21	Sucumbios	Oriente
2105	Sucumbios	21	Sucumbios	Oriente
2106	Cascales	21	Sucumbios	Oriente
2107	Cuyabeno	21	Sucumbios	Oriente
2201	Orellana	22	Orellana	Oriente
2202	Aguarico	22	Orellana	Oriente
2203	La Joya De Los Sachas	22	Orellana	Oriente
2204	Loreto	22	Orellana	Oriente

Figure D.2. **Arable lands, Ecuador**

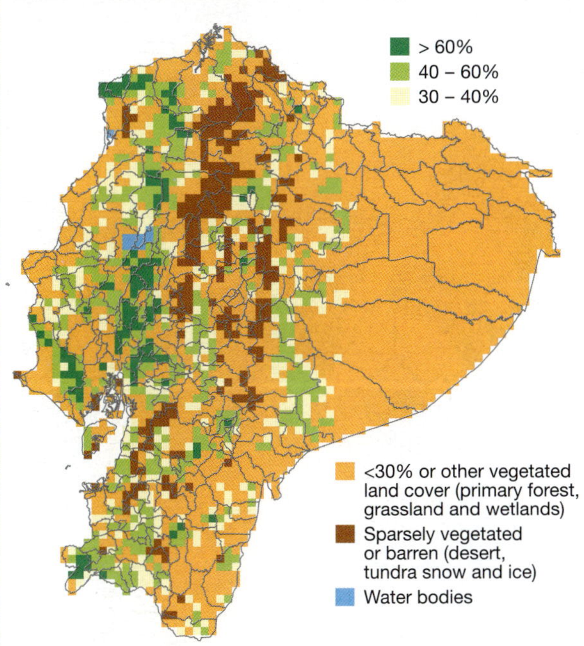

Figure D.3. **Cereal production coefficient, Ecuador (Ton/ha with coefficient)**

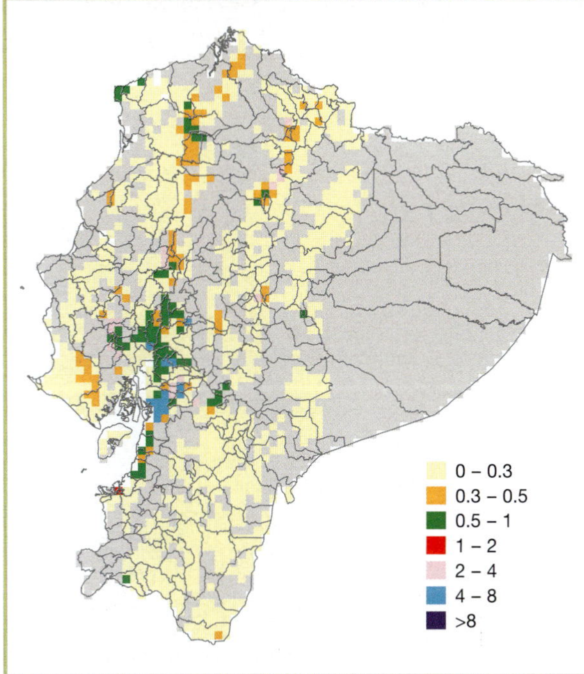

Figure D.4. **Closed forest areas, Ecuador**

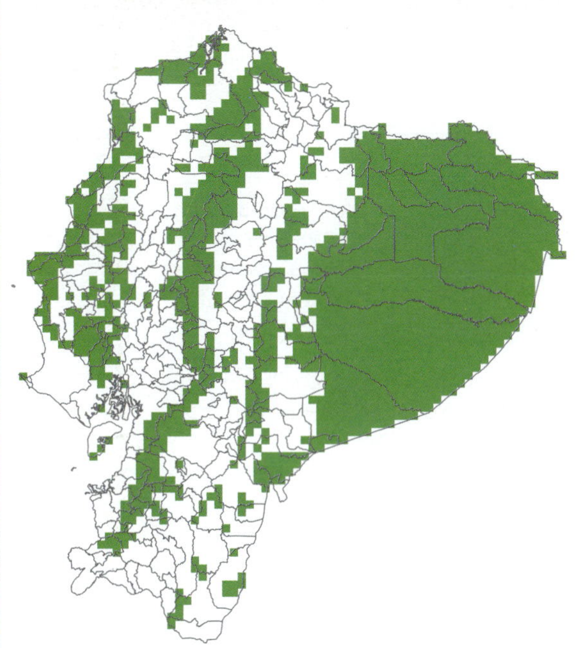

Figure D.5. **Irrigation areas, Ecuador**

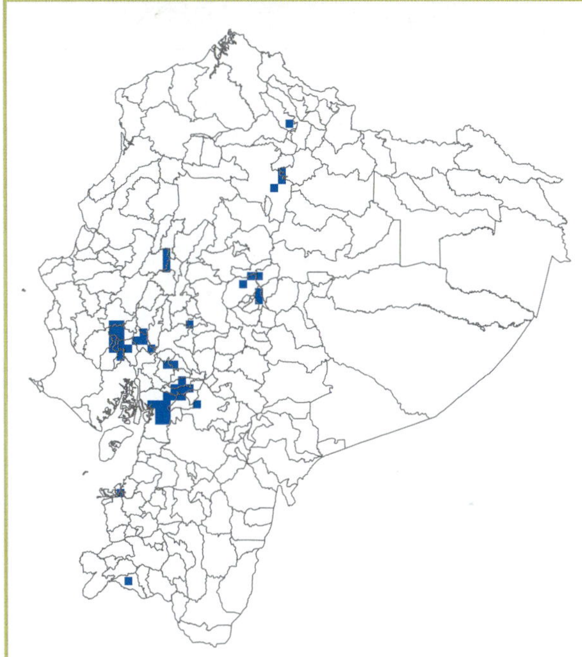

Figure D.6. **Distances from main roads, Ecuador**

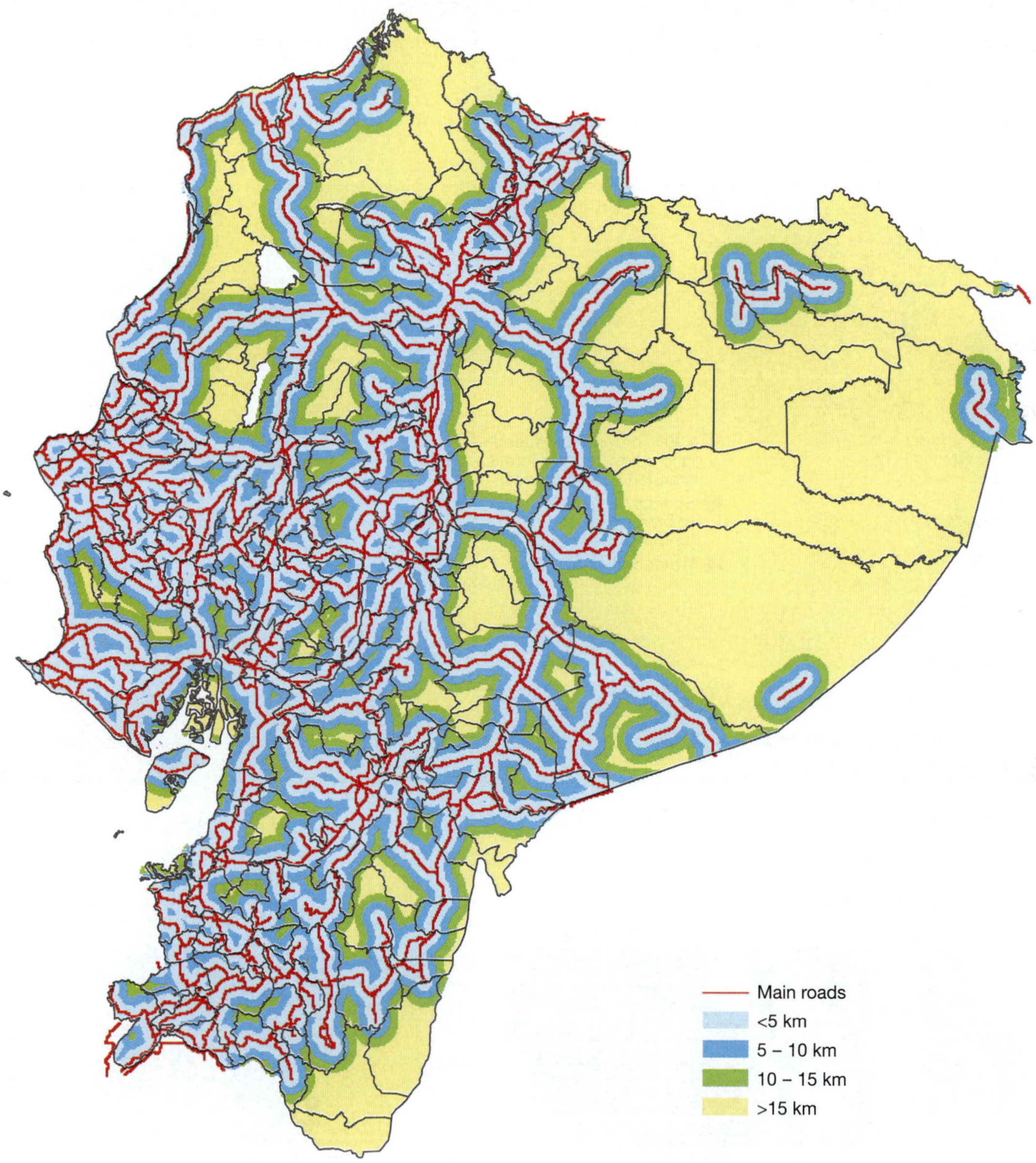

FAO Environment and Natural Resources Series

1. Africover: Specifications for geometry and cartography, 2000 (E)
2. Terrestrial Carbon Observation: The Ottawa assessment of requirements, status and next steps, 2002 (E)
3. Terrestrial Carbon Observation: The Rio de Janeiro recommendations for terrestrial and atmospheric measurements, 2002 (E)
4. Organic agriculture: Environment and food security (E)
5. Terrestrial Carbon Observation: The Frascati report on *in situ* carbon data and information (E)
6. The Clean Development Mechanism: Implications for energy and sustainable agriculture and rural development projects (E)
7. The application of a spatial regression model to the analysis and mapping of poverty (E)

Availability: May 2003

Ar	Arabic	Multil	Multilingual
C	Chinese	*	Out of print
E	English	**	In preparation
F	French		
P	Portuguese		
S	Spanish		

The FAO Technical Papers are available through the authorized FAO Sales Agents or directly from Sales and Marketing Group, FAO, Viale delle Terme di Caracalla, 00100 Rome, Italy.